Ideas

that made your

Smartphone

Why the patent system

should be abolished

Richard E. Haskell

Copyright © 2019 Richard E. Haskell
All rights reserved.

ISBN: 978-1975681739

Cover photo by Richard E. Haskell
Published by Richard E. Haskell, Inc.
Printed by CreateSpace, An Amazon.com Company

Preface

Thousands of ideas, going back 200 years, went into making your smartphone. What were some of the most important of these ideas, and who came up with them? In this book, you will meet some of these inventors and learn about their interactions with the patent system. You will see that in many cases the patent system fails to live up to its intended purpose, which is "to promote the progress of science and the useful arts." You will see that by its very nature, it does just the opposite. You will see that countless hours and dollars are wasted in the litigation of patent claims, that patents are not always awarded to the true inventor, that judges and juries are not competent to judge the validity of patents, that inventions are always based on the work of others, and that patents stifle economic growth. You will see that the very idea of regarding intellectual ideas as property is misplaced, leading one to question the very existence of the patent system.

Richard E. Haskell

Table of Contents

1. Countless Ideas   1
2. A Sealed Idea   5
3. The Forgotten Inventor   9
4. What to Talk About   12
5. Inventing Instant Communication   15
6. A Stolen Idea   21
7. Telegraphy Without Wires   25
8. Who is *The Father of Radio*?   29
9. Who Invented the Computer?   34
10. Squeezing a Computer into a Smartphone   37
11. Instant Photography Ideas   41
12. What Does a Smartphone Look Like?   44
13. What Makes a Smartphone Smart?   50
14. Storing Lots of Photos on Your Smartphone   54
15. Using Your Smartphone as a Flashlight   59
16. Who Invented the Internet?   62
17. Who Invented the World Wide Web?   65
18. What Makes a Phone Mobile?   68
19. What's with Wi-Fi and Bluetooth?   72
20. How Does Your Smartphone Know Where It Is?   75
21. Talking to Your Smartphone   78
22. Keeping Your Information Secret   81
23. Shopping on Your Smartphone   85
24. A World Without Patents   87

Chapter 1

Countless Ideas

Our riches did not come from piling brick on brick, or bachelor's degree on bachelor's degree, or bank balance on bank balance, but from piling idea on idea.

Deirdre Nansen McCloskey

In her massive book, *Bourgeois Equality—How Ideas, Not Capital or Institutions, Enriched the World*, **Deirdre Nansen McCloskey** makes the compelling case that what she calls the Great Enrichment—the unprecedented and gigantic improvement in the standards of living for everyone throughout the world, which has occurred since 1800—is the result of ideas, nurtured by liberty and dignity for commoners, and not by the accumulation of capital or the establishments of institutions.

No better example of the power of ideas is the smartphone, and its distant cousin, the cell phone, used every day by seven billion of the eight billion people on earth. The speed with which this massive transformation of instant world-wide communication took place is truly remarkable. My daughter flies to Paris, takes a selfie in front of the Fifel Tower, and immediately texts it to me. Better still, she can FaceTime with me from half-way around the world—all in living color.

The smartphone started out as just a wireless telephone but has now become the most widely-used camera in the world, taking more photos in one year (estimated to be well over three trillion) than all of the photos ever taken on film in the entire history of film cameras.

Most people probably take their smartphones for granted—something to get frustrated with when it is not working properly. But imagine what your grandparents would think of today's smartphones. It is a world they couldn't imagine. How did this happen? What were some of the big ideas that went into making it all possible? And were there institutional barriers that slowed down its development?

In the chapters to follow, we will explore some of the scientific discoveries and technological inventions that made your smartphone possible. We will consider the motivations that make scientists pursue new knowledge, make inventors invent, make entrepreneurs start new companies, all necessary ingredients for you to be able to tap a small screen held in your hand, and send a photograph half-way around the world—just like magic!

You may think you know why scientists, engineers, inventors, and entrepreneurs do what they do—money, fame, recognition, satisfaction. These undoubtedly play a role, but there is more to it than that. If you were to ask many scientists, engineers, inventors, and entrepreneurs why they do what they do, most will say, *because it is fun*! Many can't believe that they are paid to have so much fun.

The fun parts of their jobs are doing the things they love to do: detailed mathematical calculations, thinking up new ideas, conducting experiments, designing something new that no one else has ever done before, manufacturing new products, getting people to buy their product. The list of fun things to do is long. Unfortunately, so is the list of un-fun things, including spending time trying to get a patent attorney to understand your invention. But, you say, patents are important, and so it is time well spent. In this and future chapters, I will argue "no they aren't" and "no it isn't." I will make the case that the entire patent system should be abolished!

Let's begin by asking why we have patents and why you think they are important. As far as United States patents are concerned, the answer is in Article I, Section 8 of the U.S. Constitution, which states, in part,

> "The Congress shall have Power ... To promote the Progress of Science and useful Arts, by securing for limited Times to Authors and Inventors the exclusive Right to their respective Writings and Discoveries."

So, the purpose is "to promote the Progress of Science and useful Arts." We've now had over 200 years to see if patents have lived up to their noble purpose. We will look at many specific examples in future chapters, but it doesn't take much thought to realize that the very idea of a patent is not likely "to promote the Progress of Science and useful Arts," but rather to have exactly the opposite effect. The reason for this is stated boldly on the U.S. Patent Office website, where under the heading *What is a Patent?* you read the following:

> "The right conferred by the patent grant is, in the language of the statute and of the grant itself, "*the right to exclude others from making, using, offering for sale, or selling*" the invention in the United States or "importing" the invention into the United States. *What is granted is not the right to make, use, offer for sale, sell or import, but the right to exclude others from making, using, offering for sale, selling or importing the invention.* Once a patent is issued, the patentee must enforce the patent without aid of the USPTO. [*italics* added].

So, there you have it, a legal monopoly. Don't spend your time improving your product, inventing a new one, satisfying new customer needs, thinking up new needs the customer is unaware of; no, spend your time in court, fighting off presumed encroachers of your patent, the validity of which may be up to a judge or jury with little knowledge of the underlying technical issues. All at a huge cost of time and money.

This point was made with respect to biotechnology firms in an August 23, 2016 column in the Wall Street Journal by Alex Berezow and Neal Mody. They cite cases in which the U.S. Court of Appeals sometimes invalidated a patent and other times upheld a patent with little regard for the underlying science. However, their solution that "Congress ought to create a court that focuses specifically on patents that involve extremely complicated technology…staffed with judges and experts who have backgrounds in software, artificial intelligence, nanotechnology, biotechnology and other scientific fields" is completely off the mark. Putting aside the problem of finding such experts willing to waste their time going through the minutia of what was written when, the kind of thing that eats up hours of time in the current patent legal system, such experts would make a much bigger contribution to society by using their skills to move science forward, to develop new technologies, to invent new devices and processes, to start new companies, to create new jobs, in short, to do something useful.

In this book, we will ask the question, "Did the patent system really help in the development of the electric motor, the telegraph, the telephone, wireless telegraphy, radio, television, the camera, and the computer?" Your smartphone uses descendants of all these inventions. But did patent fights slow down progress? Was precious time lost? Did the five-year patent battle between Kodak and Polaroid divert their attention from the growing challenge from digital photography, which ended up sending both companies into bankruptcy?

We will pay attention to the differences between scientists, whose goal is to understand how nature works, and inventors, whose goal is to produce some useful device or product. No inventor would be able to invent anything without relying on the work of the scientists. Both scientists and inventors want recognition for their achievements, scientists by publishing their results, inventors by obtaining a patent. But do patents always reward the actual inventor? We will see that inventions are slippery things, seldom the result of the work of any single person. In fact, we will see that the central idea in any patent is always the cumulative result of countless previous ideas of others, without which the new idea would be impossible. We must ask the question: does it makes any sense to assign property rights to each such incremental new idea?

We will also question other conventional wisdom related to patents. Are patents really the incentive that causes engineers and scientists to invent? With the vast majority of patents—over 90%—assigned to someone other than the inventor (usually the inventor's employer), engineers and scientists are motivated by a lot more than patents. Do patents really stop others from competing by producing similar products? Are patents really overrated? We will address all of these questions in the following chapters.

So, who are some of the scientists and inventors whose ideas have led to the smartphone? In the following chapters, you will meet a few of the many thousands of scientists and engineers whose work has given you the smartphone you use today. The most basic thing about a smartphone is that it communicates using electromagnetic waves. Who first came up with that idea?

Chapter 2

A Sealed Idea

> *These views I wish to work out experimentally; but as much of my time is engaged in the duties of my office, and as the experiments will therefore be prolonged, and may in their course be subject to the observations of others, I wish, by depositing this paper in the care of the Royal Society, to take possession as it were of a certain date; and so have right, if they are confirmed by experiment, to claim credit for the views at that date; at which time as far as I know, no one is conscious of or can claim them but myself.*
>
> Michael Faraday
> 12 March 1832

On November 30, 1937, **Sir William Bragg**, in his role as president of the Royal Society of London, addressed its annual meeting. As was customary, he began his address by mentioning the Fellows of the Society who had died during the previous year. He then went on to other matters, including the following remarkable statement.

> On a very interesting occasion during the year, a number of letters, which had long remained under seal in the Society's safe, were opened in the presence of Fellows and others. There were no inscriptions governing the manner and time of their opening, and it seemed to the Council that any limits must long have been passed. Several of them were very interesting in their matter, quite apart from the dramatic circumstances of their disclosure so long after their authors had written and sealed them and handed them for safe keeping to the Officers of the Society. The most interesting of all was a letter of Faraday's, written in 1832; it has been published in the Society's *Occasional Notices*, 2. It shows that he was then entertaining the idea that the "diffusion" of magnetic and electric forces was comparable with the spread of waves on water or in air, and had indeed a vibratory character. Perhaps because of preoccupations which hindered the development of the idea, and because of the difficulties of

experiment, it was not until 1846 that he made public reference to the matter. In the latter year, Faraday had to take Wheatstone's place on very short notice at a Friday Evening Discourse at the Royal Institution and found that when he had completed the description of the apparatus which he and Wheatstone had arranged during the afternoon, he had still in hand a portion of the lecture hour. He then went on to speak of his "Thoughts on Ray Vibrations" which afterwards formed the subject of a paper in the *Philosophical Magazine*. It was this paper that led Maxwell *(Phil. Trans.* 1865) to put Faraday's experimental results in a mathematical form, and finally to frame the electromagnetic theory of light. Until this letter was opened at the Royal Society meeting there had been no indication that Faraday's thoughts on ray vibrations had been simmering for so long.[1]

So, there it was. **Michael Faraday**, born the son of a blacksmith on September 22, 1791, growing up in a poor area of London, would, by writing his secret sealed letter, and depositing it with the Royal Society in 1832, verify that he was the first person to conceive of the idea that there may be waves associated with electric and magnetic fields. Faraday became an apprentice bookbinder at the age of thirteen and taught himself by reading the books he was binding, never learning any mathematics. He always wished for a life in science, and finally, in 1813, was offered a job as assistant to **Sir Humphry Davy**, the director of the chemical laboratory at the Royal Institution. He accompanied Davy on an eighteen-month scientific tour of Europe, meeting many important scientists before returning to the Royal Institution to conduct experiments in chemistry and electricity. It would be fourteen years after depositing his sealed letter with the Royal Society before he spoke of these ideas publicly, in an impromptu addition to the Royal Institution's Friday evening lecture on April 3, 1846. **Charles Wheatstone** was scheduled to give the lecture, but he was very shy, hated speaking in public, and panicked at the last minute. Faraday had to step in and make the presentation, but finishing before the hour was up, decided to share his vision of what he called *lines of force*. These lines of force were not only magnetic, which he had talked about before, but also electric, and possibly gravitational. He pictured these lines of force extending throughout the universe, coming together on the bodies that we perceive to be matter. He suggested that these lines of force, when disturbed, could produce lateral vibrations, sending waves of energy along their lengths.

---

[1] Proceedings of the Royal Society, 07 December 1937, Volume 124, issue 836, page 385.

He thought that light was such a vibration of his lines of force, rejecting the idea that light must be propagating through an imponderable medium called the ether.

This Friday evening lecture about light being vibratory waves of his electric and magnetic lines of force occurred eighteen years before **James Clerk Maxwell** turned Faraday's ideas into a self-consistent mathematical form, equations that predicted the existence of electromagnetic waves travelling at the speed of light. It would be another twenty-four years before **Heinrich Hertz** first produced and detected electromagnetic waves in the laboratory. By this time, Faraday had been dead for twenty-one years and Maxwell had been dead for nine years.

Faraday was a man of science. Turning his scientific discoveries into practical applications didn't interest him. He would leave that to others. Two of his discoveries are particularly noteworthy. The first was his discovery of the law of electromagnetic induction. This law—later to be called *Faraday's law*—states that a changing magnetic field produces an electric field. Ever since **Hans Christian Oersted** showed in 1820 that an electric current in a wire would deflect a compass needle placed above and below the wire, scientists knew that electricity, in the form of an electric current, could produce magnetism—illustrated by the deflection of a compass needle. They quickly learned that by wrapping wire in a helical coil, and running current through this coil, they could produce the equivalent of a magnet—an electro-magnet. It seemed natural that a strong magnet, in the form of an electro-magnet, should be able to produce electricity. Many tried, including **Joseph Henry** in America, who made very large electro-magnets, but all failed to detect any electricity generated by the electro-magnets.

In 1831, Faraday conceived the idea of wrapping a large number of turns of wire around one side of a wrought-iron ring, about six inches in diameter, and then wrapping a second coil of wire around the opposite side of the ring. His idea was to run current through the first coil, producing a magnetic field that would go around the ring and through the second coil. His hope was that some type of vibration or wave created by the first coil would induce a current in the second coil that he could measure with a galvanometer connected to the wire of the second coil. When he ran the experiment by closing the switch connecting the battery to the first coil, he noticed a small deflection of the galvanometer needle, but then the needle returned to the zero position, indicating no steady current through the second coil. If he reversed the polarity of the battery, the galvanometer needle deflected slightly in the opposite direction, returning once more to the zero position. What Faraday had discovered was not that a large magnetic field produces an electric field (current), but rather it is a *changing* magnetic field that produces an electric field.

This may seem odd, because Oersted had shown that a *steady* current produced a magnetic field. It would be James Clerk Maxwell, thirty years later, who would show theoretically that there is another type of current—he called it a *displacement current*—that arises from a changing electric field. This means that a changing electric field produces a magnetic field, just as a changing magnetic field produces an electric field. In fact, it is just this alternation between a changing electric field producing a changing magnetic field, and vice versa, that gives rise to electromagnetic waves, making it possible for your smartphone to communicate with any other smartphone in the world!

Even if Faraday wanted to, he would not have been able to obtain a patent on his discovery of electromagnetic induction. Laws of nature and abstract ideas are not patentable. But without these discoveries of the laws of nature, there would be no electric motor, no telegraph, no telephone, no radio, no television, and no smartphone. Faraday didn't need patents to make him famous. Publishing the results of his experiments in the open literature benefited mankind, securing his place in history.

Faraday's second noteworthy discovery occurred in 1821. Starting with a permanent magnet in the center of a basin, sticking out of a pool of mercury, Faraday connected one end of a short stiff wire to a pivot point above the magnet, and inserted the other end in the pool of mercury. When he connected a battery between the pivot point and the mercury, the wire rotated about the magnet rapidly. Faraday had invented the first electric motor. He would leave it to others to turn the electric motor into a commercial success. One who tried was **Thomas Davenport**, an obscure blacksmith from Brandon, Vermont, who obtained the first U.S. Patent for an electric motor in 1837.

Chapter 3

The Forgotten Inventor

> *The discovery here claimed, and desired to be secured by Letters Patent, consist in—Applying magnetic and electro-magnetic power as a moving principle for machinery in the manner above described, or in any other substantially the same in principle.*
>
> Claim from 1837 Patent of Thomas Davenport

As mentioned in the previous chapter, Faraday succeeded in producing rotary motion of a current-carrying needle in a magnetic field in 1821. However, it would be years before others produced practical electric motors. The first electric motor to receive a U.S. patent was invented by an obscure blacksmith in Brandon, Vermont named **Thomas Davenport**. Having heard that electromagnets were being used by the Penfield Iron Works near Crown Point, NY to extract iron from pulverized ore, Davenport traveled there and bought one of the electromagnets. He brought it back to Vermont, took it apart, and used it to make an electric motor that rotated continuously.

The electromagnets used to extract iron at Crown Point, NY were made by **Joseph Henry** who was teaching at the Albany Academy at the time. He had developed the world's largest electromagnets including one that could lift 3,600 pounds. Henry became a professor at Princeton University in 1833 and then became the first secretary of the Smithsonian Institution in Washington, DC where he served from 1845 until his death in 1878.

In 1835, Davenport demonstrated his working motor to **Amos Eaton**, who was Senior Professor at the Rensselaer Institute in Troy, NY. Eaton had co-founded the institute with Stephen Van Rensselaer as the Rensselaer School in 1824, making it the oldest engineering school in the country. **Stephen Van Rensselaer** paid Davenport $30.00 for the motor and gave it to the Rensselaer Institute. Thus, the oldest engineering school in the country had the first electric motor until it was apparently destroyed in a fire in 1862. In 1861, the school's name was changed to Rensselaer Polytechnic Institute (RPI). Before moving to Princeton, Joseph Henry served as an examiner at Rensselaer.

Davenport had trouble getting a patent for his motor, but after meeting with Joseph Henry at Princeton and building new models, he finally received the first U.S. patent for an electric motor, U.S. Patent No. 132, on Feb. 25, 1837.

Henry, like Faraday, was more interested in the science than in creating practical motors. In 1831, he had demonstrated an "Electro-magnetic Engine" that produced a continuous rocking motion of an electromagnetic bar magnet by a clever scheme of breaking the circuit in such a way as to alternately reverse the polarity of the magnet. However, Henry considered this device, as well as Davenport's electric motor, to be a "philosophical toy" with no practical applications other than as interesting demonstrations for his students in his lectures.

In fact, Davenport's electric motor was a commercial failure, being about forty years ahead of its time. To power his motor, Davenport used a battery of the type invented by **Alessandro Volta** in 1800, constructed by placing plates of copper and zinc alternately in a vessel of diluted acid. What was needed was a better generator of electricity. Actually, Davenport had one. He could turn his motor into a generator of direct current (DC) electricity by simply disconnecting the battery and rotating his machine mechanically, using water power or steam.

However, it wasn't until 1882 that **Thomas Edison** used such DC generators, called dynamos, powered by coal-fired steam engines, to provide DC electricity to his incandescent lighting system in downtown Manhattan. Such DC transmission systems gave way to alternating current (AC) electrical systems invented by **Nikola Tesla** and developed by **George Westinghouse**, culminating in the use of falling water at Niagara Falls to power gigantic AC generators in 1897. The use of AC electricity was made practical due to Tesla's invention of an AC motor.

So, Davenport's motor patent turned out to be useless—an example of an invention ahead of its time. Were the patent ever to have been challenged, the courts would undoubtedly have found the single claim of the patent, given at the beginning of this chapter, to be far too broad. To claim to have invented the application of "magnetic and electro-magnetic power as a moving principle for machinery" could cover a multitude of patents to be granted over the coming decades. Indeed, in Chapter 5 we will see that in 1853 the Supreme Court rejected a similarly broad claim that tried to patent basic scientific principles.

Davenport died dejected in 1851 without fulfilling his dream of electric motors being used for running trains. However, a modern version of his DC motor, a very tiny motor, is in most smartphones today. Mounted on the shaft of the motor is an eccentric piece of metal—typically half of a circular disc. When the motor spins rapidly, your phone shakes, creating the vibrate mode.

The only recognition of Davenport's invention of the electric motor is a small plaque along the side of the road in Brandon, VT, on which is the following inscription:

> IN MEMORY OF
> **THOMAS DAVENPORT**
> 1802 – 1851
>
> ———
>
> THE INVENTOR OF THE
> ELECTRIC MOTOR
>
> ———
>
> NEAR THIS SPOT STOOD THE
> BUILDING WHERE HE DEVELOPED
> HIS INVENTION
>
> ———
>
> THIS TABLET IS PLACED HERE BY
> ALLIED ELECTRICAL ASSOCIATIONS
> IN AMERICA, IN RECOGNITION OF THE
> GREAT SERVICE RENDERED MANKIND
> BY THE INVENTION, TO THE DEVELOPMENT
> OF WHICH HE DEVOTED HIS LIFE.
> ERECTED SEPT. 28, 1910

We saw in Chapter 2 that James Clerk Maxwell provided the theoretical and mathematical foundation for Michael Faraday's idea of electromagnetic waves. But Maxwell's interests and talents were many, and as we'll see next, one of his totally different ideas has found its way into your smartphone.

Chapter 4

What to Talk About

> *From a long view of the history of mankind—seen from, say, ten thousand years from now—there can be little doubt that the most significant event of the 19th century will be judged as Maxwell's discovery of the laws of electrodynamics. The American Civil War will pale into provincial insignificance in comparison with this important scientific event of the same decade.*
>
> Richard P. Feynman

Thirty-year-old **James Clerk Maxwell** was invited to give one of the Friday Evening Discourses at the Royal Institution on 17 May 1861. He had moved to London the previous year to take up the position of professor of natural philosophy at King's College, having left a similar position at Marischal College in Aberdeen, Scotland when his position was eliminated following a merger of Merischal College with nearby King's College to form the University of Aberdeen.

What could he talk about in this Friday evening lecture? **Michael Faraday**, then sixty-nine and still Director of the Laboratory of the Royal Institution, probably invited Maxwell to give this Friday Evening Discourse. Faraday was then living in a house at Hampton Court, provided to him by the Queen on the recommendation of Prince Albert, Faraday having moved there with his wife Sarah in 1858 from their flat above his laboratory at the Royal Institution. He still visited the laboratory regularly, and most likely met with Maxwell there on occasion.

When Maxwell graduated from Cambridge University in 1854, he stayed on as a Fellow at Trinity College. During this time, he completed his first paper on electricity, called *On Faraday's Lines of Force*, which he read to the Cambridge Philosophical Society on 10 December 1855 and 11 February 1856. This paper was limited to static electric and magnetic fields, its analogy to steady flow of an incompressible fluid through imaginary tubes filling all of space making it an unlikely topic for Maxwell's Friday evening lecture.

The same unsuitability for his Friday evening lecture would probably also apply to his just-published paper called *On Physical Lines of Force*, which explained Faraday's law of electromagnetic induction by considering a new mechanical analogy involving rotating molecular vortices. Although this theory predicted wave motion at the speed of light, the implications of that important result would probably be lost on the audience.

It would be over three years in the future before Maxwell published his final theory of the electromagnetic field in a paper entitled, *A Dynamical Theory of the Electromagnetic Field*. In this paper, Maxwell dispensed with all mechanical analogies, and described the dynamics of the electric and magnetic fields themselves. These equations again predicted the existence of electromagnetic waves travelling at the speed of light. But this would be in the future, so Maxwell turned to something other than electricity to talk about in his Friday evening lecture.

He could have talked about his solution to the problem of explaining the rings of Saturn. When he was at Marischal College he entered the competition for the Adam's Prize, sponsored by St. John's College, Cambridge. That year's problem was to determine if the rings of Saturn would be stable if they were solid, fluid, or composed of many separate pieces of matter. Maxwell was able to prove that solid or fluid rings would eventually break up, and thus the rings must be made up of many separate objects. The problem was so difficult that Maxwell's was the only entry, and thus he won the prize. But he decided not to talk about this in his Friday evening lecture.

Another choice could have been his kinetic theory of gases, in which he derived the distribution of speeds of molecules making up a gas. This theory would be picked up by **Ludwig Boltzmann** and would lead to the development of the field of statistical mechanics. But instead of any of these topics, Maxwell decided to talk about one of his favorite topics, one that he had been intrigued with since he was a child, and had been working on, off and on, since he was a student at Cambridge. And so it was that on this Friday evening, Maxwell would give a lecture on his theory of color vision.

In 1855, Maxwell had published a paper entitled, *Experiments on Colour, as Perceived by the Eye, with Remarks on Colour-Blindness* in the *Transactions of the Royal Society of Edinburgh*. In this paper, he described experiments he carried out with a color wheel that he made while a student at Cambridge. The color wheel was in the form of a top that he could spin rapidly. Mounted on the axis of the top were three discs made of different colored paper—one red, one green, and one blue. Each disc had a radial slit, cut from the circumference to the center hole in the disc, allowing the three discs to be rotated relative to one another to expose various wedge-shaped areas of each color. The percentages of each color making up the resulting compound disc could be read easily from markings that Maxwell had placed on a ring around the circumference of the discs. He added a fourth smaller disk on top of the three colored discs, the smaller disc containing wedges of white and black to form a neutral gray when the top was spun. Maxwell would adjust the percentages of red, green, and blue until his subjects, other students at Cambridge, would agree that the outer color of the spinning top matched the gray of the central disc. He also could change the color of the inner disc so as to match colors other than gray. Maxwell invented a color triangle to describe any color in terms of the amounts of red, blue, and green.

In the same 1855 paper in which he introduced his color wheel, Maxwell described how to make a color photograph by taking three different photographs through red, green, and blue glass plates, and then projecting them through three magic lanterns containing the same red, green, and blue glass plates onto a screen. Inasmuch as the audience at his Friday evening lecture would not be able to see his color wheel, Maxwell decided to try to make a color photograph that he could project onto a screen for the entire audience to view.

With the help of the photography expert **Thomas Sutton**, Maxwell took three separate photographs of a tartan ribbon through red, green, and blue glass plates. At his Friday evening lecture at the Royal Institution, Maxwell used the same three glass plates to project the three photographs through three magic lanterns onto a screen at the front of the auditorium. The audience at Maxwell's lecture on 17 May 1861 viewed the world's first color photograph.

Maxwell's idea of mixing various amounts of red, green, and blue colors to produce any color is at the heart of all methods of viewing color images, whether a color photograph, a color TV, or the color screen on your smartphone.

At the time that Faraday was discovering the law of electromagnetic induction in 1831, the time to communicate from one location on earth to a distant location was determined by the speed of a horse or a ship. (The railroad boom was about to start.) However, a revolution in nearly instant communication was about to begin with the invention of the electric telegraph, in England by **William Cooke** and **Charles Wheatstone,** and in the United States by the painter, **Samuel Morse**.

Chapter 5

Inventing Instant Communication

> *8. I do not propose to limit myself to the specific machinery or parts of machinery described in the foregoing specifications and claims, the essence of my invention being the use of the motive power of the electric or galvanic current, which I call "electro-magnetism," however developed, for marking or printing intelligible characters, signs, or letters at any distances, being a new application of that power of which I claim to be the first inventor or discoverer.*
>
> Claim 8 from 1848 Patent of Samuel F. B. Morse

Samuel Finley Breese Morse was an unlikely candidate to be the inventor of the electric telegraph. Born in Charlestown, Massachusetts on April 27, 1791 to a minister father and a mother who was the granddaughter of the president of Princeton University, Finley, as his family called him, became an accomplished artist, painting portraits of such people as John Adams, James Monroe, Eli Whitney, and a full-length painting of the Marquis de Lafayette. After graduating from Yale in 1810 with a dream of becoming a painter of historical paintings, he painted *The Landing of the Pilgrims at Plymouth* before heading off to London to sharpen his skills over the next four years. While in London he painted *Dying Hercules*, an eight-foot by six-foot painting that he brought home and tried to exhibit or sell without success.

In August 1816, Morse met seventeen-year-old Lucretia Pickering Walker, becoming engaged a month later, and marrying her two years after that. To make ends meet, Morse traveled a lot, painting portraits, but still dreaming of painting an important historical painting. In 1823, he painted his *House of Representatives*, a large eleven-foot by seven-and-a-half-foot painting showing all of the members on the House floor, plus a couple of extra people in the balcony, including his father. Attempts to raise money by charging the public to view the painting mostly failed.

By the end of 1824, Lucretia was living in New Haven, Connecticut where Morse's father had moved five years earlier. By this time, Morse and Lucretia had two children, five-year-old Susan and one-year-old Charles, with another on the way. In early 1825, Morse made a quick stop in New Haven a couple of weeks after the birth of his third child James, before heading off to Washington to start a portrait of Lafayette.

In Washington, he met Lafayette and got to witness the excitement in the House of Representatives to resolve the presidential election between **John Quincy Adams** and **Senator Andrew Jackson**. Jackson had won the popular vote but not a majority of electoral votes, the House eventually awarding the presidency to Adams.

Morse wrote to Lucretia describing all the exciting news in Washington, but received no reply. Eventually, he received a letter from his father, telling him that Lucretia had died about a week after Morse had left for Washington. Morse was obviously distraught, even having missed the funeral. One wonders if this long delay in learning about the death of his wife might have instilled in the back of his mind the need for more instant communication.

Morse moved to New York City where he continued to paint and become involved in politics. In 1826, he established the National Academy of the Arts of Design, where young artists could meet, practice, and receive instruction in painting, sculpture, architecture, and engraving. He ran for mayor of New York City on an anti-immigration and anti-Catholic platform, coming in a distant last in a field of four candidates.

In late 1829, Morse began a three-year trip to Europe, spending time in Rome, Florence, and Paris. While in Paris he painted *The Grand Gallery of the Louvre*, a six-foot by nine-foot canvas containing dozens of miniature copies of Louvre masterpieces. Morse included himself in the painting as well as **James Fennimore Cooper**, whom he had befriended in Paris, together with Cooper's wife and daughter.

Morse began his return voyage from France aboard the ship *Sully* in October 1832. While dining with a fellow passenger, **Dr. Charles Jackson**, a physician-geologist from Boston, the topic of recent discoveries in electro-magnetism was discussed. Morse recalled that it was during these discussions that he first conceived the idea of using electricity to send intelligent messages long distances over wires.

When he returned to New York, Morse resumed his painting and his affiliation with the National Academy of Design. However, over the next five years, he also secretly built an electric telegraph system in his apartment. The apparatus consisted of a transmitting device he called a *port-rule* and a receiving device he called a *register*.

A newspaper article in the spring of 1837 reported that a pair of Frenchmen claimed to have invented a new telegraph system for long-distance communication and planned to demonstrate it in the United States. Concerned that he might lose credit for inventing his telegraph system, Morse finally went public with his telegraph apparatus. As it turned out, the French system was simply an enhanced semaphore system, of the type already installed throughout France, consisting of movable arms on hill-top towers, separated by several miles, and viewed through a telescope by an operator who would pass on the message to the next tower.

In Morse's initial electric telegraph system, the transmitting operator would set metal blanks on the port-rule in such a way as to encode a number. Turning a crank would cause these metal blanks to move under a lever, making and breaking an electric circuit, sending the coded signal along a wire to the receiving device. This receiving device, consisting of a pendulum, gears, and a moving paper roll, would cause the coded number to be printed with a pencil as a sequence of dashes. The receiving operator would look up a word in a dictionary corresponding to a particular number.

Later, Morse simplified his system by introducing his famous Morse code, representing individual letters of the alphabet as sequences of dots and dashes. To make his telegraph practical for long-distance communication, Morse invented an electromagnetic relay that would allow a new battery to be inserted in the telegraph line every twenty miles or so.

For the next three years, Morse improved his telegraph system, and took on partners, many of whom would cause him grief in the ensuing years. He travelled to Europe where he was unsuccessful in obtaining a British patent and failed to gain any financial support. While in Paris in early 1839, he met **Louis Jacques Mandeé Daguerre**, who had developed a new process for recording permanent photographic images on thin metal sheets. A camera obscura was a dark box with a lens in one end, which could project an outside image onto a viewing surface, a device often used as an aid by painters. Daguerre replaced the viewing surface with a thin metal plate sensitized with iodine vapor, followed by exposing the plate to the image for up to several minutes. Daguerre discovered that the latent image could be made visible by exposing the plate to hot mercury vapor, and then fixing the image with sodium thiosulfate.

Morse was fascinated by these Daguerreotypes, and when he returned to New York he started using Daguerre's recipe for making these new types of photographs. By the spring of 1840, Morse had opened a portrait studio for taking Daguerreotypes of people, and even gave classes on how to make such photographs. He experimented with different techniques and with certain lighting could reduce the exposure time to less than a minute. Morse would undoubtedly be stunned to see how trivial it is today for anyone to take color photographs or even video clips with their smartphone.

Morse filed his first U.S. telegraph patent in 1840 (Patent No. 1,647) and tried for nearly three years to get the U.S. Congress to pass a $30,000 appropriation to fund the building of a test telegraph line between Baltimore and Washington, DC. The bill finally passed in early 1843, and on May 24, 1844, Morse tapped the message "What hath God wrought," sending it over the telegraph line from the Supreme Court chamber in the Capitol Building to Baltimore, where his partner, **Alfred Vail**, tapped the message back to him. While Morse took credit for inventing the telegraph, it was Alfred Vail who provided much of the inventiveness and mechanical skill to get everything to work. This is not uncommon. The names of the inventors on a patent often do not include those most responsible for making the invention work.

Following this successful message, Morse attempted to interest the government in purchasing and developing his electric telegraph. When this attempt failed, he formed a business relationship with **Amos Kendall**, who set up the Magnetic Telegraph Company in May 1845 to complete the Washington-Baltimore line through Philadelphia to New York. Kendall handled all the business dealings, forming various companies to build commercial telegraph lines in different parts of the country. By the fall of 1846, telegraph messages could be sent between Washington, New York, Boston, Philadelphia, and Buffalo. Lines were being built from Philadelphia through Pittsburg and Cincinnati to St. Louis and from Buffalo through Cleveland and Detroit to Chicago.

All this activity attracted those looking to cash in on potential profits, and inevitably required Morse to defend his patent claims in court. A total of fifteen litigations over the next six years culminated in the *Great Telegraph Case* before the U.S. Supreme Court beginning in December 1852. Morse had sued **Henry O'Reilly**, who Morse had licensed to build a telegraph line as far as St. Louis. But O'Reilly decided to start his own company and continue the line to New Orleans and eventually all the way to California. O'Reilly argued that his telegraph line used improved technology not covered by Morse's patent, but O'Reilly lost in a Kentucky circuit court judgment.

The seven Supreme Court justices, led by **Chief Justice Roger B. Taney**, listened to O'Reilly's counsel, **Senator Salmon P. Chase**, insist that Morse did not invent the art of telegraphing by electro-magnetism, that he did not invent the battery, the electro-magnet, the levers, or any of the other parts or processes that went into his telegraph system. It is true that all inventions build upon the ideas and inventions of others, making it a subjective call at best, by justices in this case with little technical knowledge, as to the priority due any single inventor.

After a full year of deliberation, the Supreme Court finally ruled in February 1854 that Morse was indeed the sole inventor of the electric telegraph, choosing to ignore the fact that in 1837 Cooke and Wheatstone had invented and patented a telegraph system used in England in which a letter was encoded by the positions of five needles at the receiving end. Morse argued that his system was better because it gave a permanent record on paper of his dots and dashes code. The irony of his argument is that eventually telegraph operators learned to recognize the letter codes by their sound, writing down the message as it came in.

In ruling in favor of Morse, however, the Supreme Court upheld only the first seven of Morse's claims. It struck down Claim 8 (quoted at the beginning of this chapter) as being over-broad. This claim only appears in the 1848 Reissue No. 117, which was a reissue of the patents of 1846, which was a reissue of the original patent No. 1,647 of 20 June 1840. In striking down Claim 8, Chief Justice Taney wrote:

> It is impossible to misunderstand the extent of this claim. He claims the exclusive right to every improvement where the motive power is the electric or galvanic current, and the result is the marking or printing intelligible characters, signs, or letters at a distance.
>
> If this claim can be maintained, it matters not by what process or machinery the result is accomplished. For aught that we now know, some future inventor, in the onward march of science, may discover a mode of writing or printing at a distance by means of the electric or galvanic current, without using any part of the process or combination set forth in the plaintiff's specification. His invention may be less complicated—less liable to get out of order—less expensive in construction, and in its operation. But yet if it is covered by this patent, the inventor could not use it, nor the public have the benefit of it, without the permission of this patentee.
>
> Nor is this all; while he shuts the door against inventions of other persons, the patentee would be able to avail himself of new discoveries in the properties and powers of electro-magnetism which scientific men might bring to light. For he says he does not confine his claim to the machinery or parts of machinery which he specifies, but claims for himself a monopoly in its use, however developed, for the purpose of printing at a distance. New discoveries in physical science may enable him to combine it with new agents and new elements, and by that means attain the object in a manner superior to the present process and altogether different from it. And if he can secure the exclusive use by his present patent, he may vary it with every new discovery and development of the science, and need place no description of the new manner, process, or machinery upon the records of the patent office. And when his patent expires, the public must apply to him to learn what it is. In fine, he claims an exclusive right to use a manner and process which he has not described and indeed had not invented, and therefore could not describe when he obtained his patent. The court is of opinion that the claim is too broad, and not warranted by law.

You cannot patent an idea. The idea of a telegraph system had occurred to lots of people. You cannot patent a law of nature. The laws of electromagnetism, as discovered by Ampère and Faraday, and put into mathematical form by Maxwell, are central to the operation of the electric telegraph, but are not subject to being patented. And claims such as Morse's

Claim 8 are too broad and have been rejected by the Supreme Court. This leaves the narrow region of some specific implementation, an implementation often easy to get around with some slight modification, as suitable for obtaining a patent. All inventions are based on countless previous ideas. Is it fair to credit an invention to someone who adds perhaps one or two new ideas and then claims credit for all previous ideas upon which the invention is based? Are the costs and time devoted to defending patents in a court of law worth it?

The long legal battles over his patents were not only a major distraction, but also caused Morse much anxiety. Right up until the time of his death in 1872, articles were being written attempting to correct the record of who contributed the most to the invention and development of the electric telegraph, and Morse felt compelled to rebut these charges and continue to make his case.

The source of all the anxiety experienced by many inventors is the patent system itself. The patent system was not responsible for the development of the telegraph system. Without a patent system, the telegraph would have been invented and implemented anyway, perhaps sooner. In fact, an alternate telegraph system had already been patented in England by Cooke and Wheatstone. As it was, it took Morse several years to convince people of its practicality. But once its potential and benefits became obvious, there was no shortage of investors who wanted to get in on the action.

Eventually, the Morse system became the default system worldwide, not because of his patents, but because his system was cheaper, simpler, and more reliable than competing systems. That is the secret to making an invention successful—make it cheaper, simpler, and more reliable than the competition. You don't need a patent to do that. In fact, having a patent may make you complacent, allowing someone else to make a competing product that is cheaper, simpler, and more reliable.

While the world was busy communicating by sending dots and dashes over wires, others were thinking of sending voice and music over the same wires. Many people had this idea, and sometimes the wrong person is awarded a patent.

Chapter 6

A Stolen Idea

> *Of one thing I am quite determined and that is to waste no more time and money on the telephone.... I am sick of the telephone and have done with it—excepting as a plaything to amuse my leisure moments.... There is too much of the element of speculation in patents for me.*
>
> Alexander Graham Bell
> Letter to his wife Mabel,
> September 1878

Two years after being granted the "telephone patent," which would become one of the most lucrative patents ever issued, **Alexander Graham Bell** was disgusted with his invention of the telephone and wanted nothing more to do with it. Why? In his book *The Telephone Gambit—Chasing Alexander Graham Bell's Secret*, Seth Shulman provides convincing evidence, rumored and speculated on for years, that Alexander Graham Bell stole from **Elisha Gray** the idea of using a diaphragm attached to a needle immersed in acidic water, a diaphragm that would vibrate when speaking, creating a variable resistance that could be detected as a variable voltage at the other end of the line.

This idea shows up for the first time, out of the blue, in Bell's notebook on March 8, 1876, the day after returning from Washington, DC, where an unscrupulous, alcoholic patent examiner, **Zenas Fisk Wilber**, showed Bell a sketch of the diaphragm-needle-liquid idea from a caveat submitted to the U.S. Patent Office by Elisha Gray on February 14, 1876. Ten years later, in an affidavit dated April 8, 1886, Wilber admitted that he had showed Gray's caveat to Bell, and explained to Bell how it worked.

Two days after sketching in his notebook the diaphragm-needle-liquid method of transmitting voice sound over a wire, a sketch strikingly similar to the sketch in Gray's caveat, Bell spoke into his diaphragm-needle-liquid transmitter, "Mr. Watson—Come here—I want to see you," whereupon Mr. Watson appeared from the next room, having heard the message at the receiver end of the wire. This account of the invention of the telephone is the story of legend, promoted over the years by the Bell Telephone Company.

Remarkably, Bell never used the diaphragm-needle-liquid transmitter again. When he first publicly demonstrated his telephone at the Centennial

Exposition in Philadelphia on June 25, 1876, with Elisha Gray witnessing the demonstration, he used a magneto transmitter that had never worked prior to his patent application. Bell was very reluctant to attend the exposition at all but was forced to go by his eighteen-year-old fiancée Mabel and her father, **Gardiner Greene Hubbard**, an attorney and entrepreneur, who had played a central role in getting Bell's "telephone patent" accepted ahead of Elisha Gray's caveat, and a year later would organize the Bell Telephone Company and become its first president. At this same Centennial Exposition, Elisha Gray demonstrated his "Electro-harmonic Telegraph" by playing "Home, Sweet Home" on a keyboard and transmitting the musical sounds several hundred feet over a telegraph wire.

The striking thing about Bell's "telephone patent" is that the patent is not about a telephone. The title of the patent is *Improvement in Telegraphy*, and the patent is about how to send multiple telegraphic signals on a single wire simultaneously. This was an important problem at the time to increase the throughput of telegraphic communication. In fact, Elisha Gray demonstrated his solution to the problem by sending eight messages on a single wire at the Centennial Exposition. Almost the entire content of Bell's patent is related to this problem. The only reference to using a needle immersed in a liquid to vary the resistance is in a section originally written in the margin of the handwritten application, and only the last of five claims mentions "transmitting vocal or other sounds telegraphically," almost as an afterthought. On the other hand, the title of Gray's caveat was *Instruments for Transmitting Vocal Sounds Telegraphically*.

The idea of a telephone had been around for years, and actual telephone devices had been demonstrated. The German scientist and inventor **Philipp Reis** made a working version of what he called a "telephon" in 1861, but never patented it. There is evidence that Bell was well aware of this previous work. **Antonio Santi Giuseppe Meucci**, an Italian inventor living in Staten Island, New York developed a voice-communication device linking his bedroom with his laboratory. He submitted a caveat entitled "Sound Telegraph" to the U.S. Patent Office on December 28, 1871, but never followed up with a complete patent application.

Bell's magneto transmitter that he used for the Centennial Exposition demonstration of his telephone produced only marginal results—but enough to make an impression at the exposition. Not until a "carbon microphone" was used for the telephone transmitter, independently invented by **Thomas Edison** in 1877 and earlier by **David Edward Hughes** in England, could clear vocal sounds be heard on a telephone.

The widespread belief that Alexander Graham Bell invented the telephone is typical of how history is often written by the winners, creating legends often at odds with the facts. The evidence is overwhelming that Bell stole from Elisha

Gray the idea of the liquid transmitter that he used in his first successful telephone transmission. Recall that Bell's "telephone patent" had nothing to do with a telephone, the references to a liquid transmitter and voice transmission being added to the patent application after learning the contents of Gray's caveat. The fact that he abandoned all work on developing a working telephone within two years and spent the next couple of decades in court fighting off lawsuits brought by many different people, including Elisha Gray and Thomas Edison, raises the question of whether the patent system is doing its job.

Did the patent system make a positive contribution to the development of the telephone? It's hard to see how. Arguably, the patent for the telephone was granted to the wrong person. The costs involved in litigating the 587 court cases involving challenges to telephone-related patents, including five Supreme Court decisions, one of which came within one vote of overturning the Bell patent, seem all out of proportion, and unlikely to help in "promoting the progress of science and useful arts."

Suppose there were no patent system. The telephone would have been invented anyway, probably sooner. Bell wouldn't have had to endure the years of anguish and court testimony, which turned him against the telephone. If everyone did what David Edward Hughes did in the 1870s by not seeking a patent on his carbon microphone, but rather making it available for anyone to use, then all of the best ideas for different parts of a telephone would rise to the top, and a better telephone, developed more quickly, would be the result. Companies such as Bell Telephone, Western Electric, and Western Union could then compete on an even playing field to see who could satisfy the most customers with the best telephones at the lowest cost. The public would be the biggest winners and might not have to wait over half a century to get a phone that wasn't black, or wait over a century to get a phone that they could carry in their pocket and call anywhere in the world without being connected to wires.

No one person invented the telephone—not Bell, not Gray, not anyone. It took dozens of ideas, by dozens of people, over many years to produce the first working telephone. For today's smartphone, we are talking about thousands of ideas, by thousands of people, for over a hundred years. Does granting a patent to one person for an invention based on the ideas of thousands of other people make any sense?

By the end of the 19th century, nearly half a million telephones were in use, mostly for local calls. Long distance communication was dominated by the telegraph, with over 63 million messages sent in 1900 over hundreds of thousands of miles of wire, including a transatlantic cable that was first operational in 1866.

In 1888, **Heinrich Hertz** was the first to generate and detect electromagnetic waves in the laboratory. The following year the British scientist **Oliver Lodge** improved on Hertz's method of detecting the waves. Both men were scientists, interested in learning how the world worked. There is no evidence that either of them thought that electromagnetic waves would be a practical means of long-distance communication. Who was the person who had the crazy idea that the invisible waves generated by Hertz could be used for wireless telegraphy over long distances? It was a young twenty-year-old, who learned a lot about electricity by reading books and journals as a teenager, growing up in his father's Villa just south of Bologna, Italy.

Chapter 7

Telegraphy Without Wires

> *At my home near Bologna, in Italy, I commenced early in 1895 to carry out tests and experiments with the object of determining whether it would be possible by means of Hertzian waves to transmit to a distance telegraphic signs and symbols without the aid of connecting wires.*
>
> Guglielmo Marconi
Nobel Lecture
December 11, 1909

When **Heinrich Hertz** died on New Year's Day, 1894, at the age of thirty-six, the papers were full of stories about how he was the first to generate and detect electromagnetic waves in the laboratory in 1888. The waves were generated by sparks produced by an induction coil and were detected by observing a weak spark in a small gap in a loop of wire. Hertz was able to measure a number of wave properties including the wavelength of the waves.

When twenty-year-old **Guglielmo Marconi** read these reports of how Hertz had generated and detected electromagnetic waves, he became obsessed with the idea of using such waves to send telegraphic messages over long distances without wires. Such a ludicrous idea wouldn't have occurred to Hertz or other scientists studying these Hertzian waves, including **Oliver Lodge**, who was a professor of physics and mathematics at the University College, Liverpool. After all, they knew that light was a type of electromagnetic wave, and therefore they assumed that the waves traveled in a straight line, and a detector would have to be within the line-of-sight of the transmitter. They knew that these invisible waves went through walls, and they could detect them in nearby buildings, but long distant communication of hundreds or thousands of miles, they thought, was out of the question. Marconi, of course, didn't know this was "impossible," and so he charged ahead.

Guglielmo Marconi grew up living in Villa Griffone, the family estate south of Bologna, his father Giuseppe Marconi a prosperous landowner and his Irish mother Annie Jameson, the daughter of the wealthy distiller of Jameson's Irish whiskey. As a teenager, Guglielmo read books and journals from the vast library at Villa Griffone, including Faraday's descriptions of his electrical experiments and books about Benjamin Franklin's invention of the lightning conductor. He began experimenting with electricity when he was fifteen, and at the age of eighteen was allowed to attend lectures given by **Augusto Righi**, Professor of Physics at the University of Bologna, who had also worked on Hertzian waves. Guglielmo turned a second-floor room at Villa Griffone into a laboratory for conducting his electrical experiments.

By 1895, Marconi had built an induction coil spark generator of the type that Hertz and Righi had used. Oliver Lodge had developed a new detector that he called a *coherer*, which was more sensitive than the wire loop used by Hertz. Using results first discovered by **Edouard Branly** in Paris, Lodge found that if he put metal filings in a glass tube with wires at each end, the resistance across the tube went way down when a nearby spark generated electromagnetic waves, as the metal filings tended to "cohere." Tapping the tube would return the resistance of the metal filings to the high-resistance state. Marconi used this coherer to detect his electromagnetic waves.

Marconi spent hours, day after day, improving his transmitter and receiver by trial and error. His goal now was to see how far away he could detect a signal sent from his transmitter. In the summer of 1895, he found that if he connected one terminal of his spark generator to the earth and the other terminal to a wire extended high above the ground, and if he did the same with the coherer at the receiver, he could send the signal a much greater distance, even over hills when the receiver was out of sight. He had discovered an antenna for transmitting and receiving the invisible waves, waves he did not really understand, except by one experiment after another.

In February of 1896, Marconi and his mother moved to England, where his cousin, **Colonel Henry Jameson-Davis**, a professional engineer, advised him to seek patent protection before revealing details of his wireless telegraph invention. Marconi established a relationship with **William Preece**, engineer-in-chief at the British Post Office, who initially helped Marconi demonstrate his wireless telegraph on the Salisbury Plain and then, in May 1897, conduct tests between Lavernock Point, near Cardiff, and the island of Flat Holme, over three miles away.

When Marconi's first wireless telegraph patent was issued in July 1897 (British patent no. 12039 and U.S. patent no. 586,193), his relationship with Preece and the British Post Office was severed. That same month, Marconi and his cousin Jameson-Davis founded the "Wireless Telegraph and Signal Company" to commercialize his invention. The basic components of Marconi's wireless telegraph system were a transmitter, using an induction coil spark generator, and a receiver, using a coherer. All of these components were well-known and had been invented and developed by others. But this did not stop Marconi from including all of them in the claims of his patents—a total of fifty-six claims in his U.S. patent. Marconi's one new idea, really a discovery, was that if he stuck one end of his spark generator and one end of his coherer into the ground, he could transmit over longer distances. It is true that Marconi modified and tested all of the components invented by others, and, by trial and error, achieved transmission distances far greater than anyone had imagined possible. But is it fair to claim credit for the scores of ideas of others, ideas without which your invention would be impossible?

Marconi became an almost instant celebrity, moving in the upper-echelons of society and demonstrating his system for Queen Victoria on the Isle of Wight. The distance over which Marconi could communicate increased from tens of miles to hundreds of miles until finally on December 12, 1901 he received three dots, the Morse code for the letter 'S', in St. John's, Newfoundland, sent from a transmitter in the English town of Poldhu in Cornwall, the first transatlantic wireless transmission. President Theodore Roosevelt sent a wireless message from the United State to King Edward VII in England in January 1903.

The most important application of Marconi's wireless telegraphy was in sending messages from ship to ship at sea and from ship to shore. Hundreds of lives were saved on wrecked ships because the ships were equipped with Marconi's wireless equipment and operators, the most famous example being the sinking of the Titanic in 1912.

In 1909, Marconi shared the Nobel prize in physics, not because he was a scientist, which he was not, but because his dogged development of wireless telegraphy had such a positive impact on mankind.

Could Marconi have developed his wireless telegraphy system if there were no patents? Undoubtedly so. If anything, writing the many patents he submitted and defending them in court slowed him down. Until he had succeeded, he had little competition, because most scientists didn't think his dream of long distant communication was possible. And when he did succeed, his patents didn't deter others from introducing similar systems, particularly in Germany and the United States. Fighting patent infringers is a never-ending battle, a battle that the courts are uniquely unqualified to handle.

In 1943, six years after Marconi's death, the U.S. Supreme Court heard the case of Marconi Wireless Telephone Corporation of America v. United States, and in a 5 to 3 decision, with one abstention, ruled that an important Marconi patent from 1904, related to tuning multiple transmitting and receiving stations to different frequencies, was invalid, having been anticipated by patents of Oliver Lodge in 1898 and John Stone Stone, an American mathematician, physicist and inventor, in 1902. The weirdness of deciding such a case forty years after the fact was not lost on one of the dissenters, **Justice Felix Frankfurter**, who wrote in his dissent:

> To find in 1943 that what Marconi did really did not promote the progress of science because it had been anticipated is more than a mirage of hindsight. Wireless is so unconscious a part of us, like the automobile

to the modern child, that it is almost impossible to imagine ourselves back into the time when Marconi gave to the world what, for us, is part of the order of our universe. And yet, because a judge of unusual capacity for understanding scientific matters is able to demonstrate by a process of intricate ratiocination that anyone could have drawn precisely the inferences that Marconi drew and that Stone hinted at on paper, the Court finds that Marconi's patent was invalid although nobody except Marconi did in fact draw the right inferences that were embodied into a workable boon for mankind. For me, it speaks volumes that it should have taken forty years to reveal the fatal bearing of Stone's relation to Marconi's achievement by a retrospective reading of his application to mean this, rather than that. This is, for me, and I say it with much diffidence, too easy a transition from what was not to what became.

I have little doubt, insofar as I am entitled to express an opinion, that the vast transforming forces of technology have rendered obsolete much in our patent law. For all I know, the basic assumption of our patent law may be false, and inventors and their financial backers do not need the incentive of a limited monopoly to stimulate invention. But whatever revamping our patent laws may need, it is the business of Congress to do the revamping. We have neither constitutional authority nor scientific competence for the task.

Do inventors really need the incentive of a limited monopoly in order to invent?

While Marconi was focused on sending dots and dashes long distances from his spark transmitters, others were looking for methods of sending speech and music by wireless transmission.

Chapter 8

Who is *The Father of Radio?*

> *Scientific questions raised in infringement suits are often of such difficulty that a conscientious but technically untrained judge passes upon them only in hesitation.*
>
> New York Times Editorial
> *A Needed Patent Reform*
> June 4, 1934

Ask people to name *The Father of Radio*, and the most likely reply would be "Marconi." While the wireless telegraph that Marconi did invent was a precursor to radio, it was not radio. Marconi's wireless telegraphy sent dots and dashes from one point to another. Radio broadcasts music and speech from one transmitter to a multitude of listeners, who can tune their radios to any number of different stations. So, who was *The Father of Radio*?

Could it be **John Ambrose Fleming**, a professor at University College London and consultant to Marconi, who, in 1905, patented an *Instrument of Converting Alternating Electric Currents into Continuous Currents*? This U.S. Patent No. 803,684 was the first invention of the thermionic vacuum tube that would play an indispensable role in the development of radio. Fleming got his idea from recalling the *Edison effect*—an anomaly that Thomas Edison had noticed in his electric light bulbs in 1875. Edison even took out a patent on the phenomenon, even though he did not understand why the heated carbon filament was causing the inside of his glass bulbs to turn black. Fleming's device was just a diode, in which electrons from the heated carbon filament could flow from the filament to a cold metal plate inside the tube, but not in the opposite direction. Fleming's motivation was to use the device as a more sensitive method of detecting wireless telegraphic signals than Marconi's coherer. So, if not Fleming, who else might be *The Father of Radio?*

Lee de Forest would certainly have nominated himself, and in fact, he did, even including the phrase in the title of his autobiography. Growing up in Alabama, the son of a minister, de Forest always longed to be a rich and famous inventor. After earning a Ph.D. from Yale in 1899, he worked briefly at Western Electric Co. and some other jobs, and then, in 1902, became Scientific Director of American DeForest Wireless Telegraph Co., formed to promote de Forest's work by the promoter **Abraham White**, often accused of running a fraudulent enterprise. De Forest tried to develop a more sensitive wireless telegraph receiver, using a variety of methods. On Jan. 29, 1907, he filed for a patent on *Space Telegraphy*, in which he introduced a third element, called a *grid*, into Fleming's thermionic vacuum tube. He found that the use of this tube, which he called an *Audion*, in the telegraph receiving circuit greatly increased the sensitivity. The patent, No. 879,532, was issued on Feb. 18, 1908.

The Audion was a triode, a type of vacuum tube that would become widespread in the development of radio. But de Forest didn't develop it with radio in mind, and he didn't even understand how it worked. He thought that the glass tube needed to have some residual gas in it, which would become ionized. When Bell Labs later studied the triode in detail and developed it for use in telephone systems, they discovered that a very high vacuum in the tube was required to provide consistent and better performance.

So, if de Forest didn't really know how the Audion worked, and if he didn't conduct any test radio broadcasts until 1907, did anyone broadcast entertainment over radio before him?

Reginald Fessenden, seven years older than de Forest, was an inventor with hundreds of patents, many in the field of radio technology. His 1902 patent (No. 706,747) entitled, *Apparatus for Signaling by Electromagnetic Waves*, describes a transmitting and receiving device for sending and receiving spoken words wirelessly. He had been working on similar systems for some time, and was the first to send a wireless telephone message on December 23, 1900. On Christmas Eve, 1906, Fessenden broadcast the first ever radio entertainment show, consisting of speech, singing and music, to wireless telegraph operators aboard ships along the Atlantic coast. This event was soon forgotten, Fessenden being more interested in the possibilities of wireless telephony. The idea that broadcasting the same thing to multiple users would have commercial value was still a little ahead of its time. It is often the case that things that seem obvious in retrospect, are not obvious at the time.

For radio to become practical, better methods for generating and detecting continuous waves, preferably using vacuum tubes, would need to be invented. Could the person who did that be called *The Father of Radio*?

Edwin Howard Armstrong graduated from Columbia University with an electrical engineering degree in 1913. Later that year he filed his first patent for a regenerative receiver that used feedback with de Forest's Audion to greatly increase the receiver sensitivity and could also be modified to be an oscillator that generated continuous electromagnetic waves. He later invented the superheterodyne receiver—used in all radios to tune to specific stations—as well as frequency modulation. Armstrong's inventions helped to make radio a reality, but did this make him *The Father of Radio*? Someone needed the vision to see broadcasting as a potential commercial success?

David Sarnoff, born in Russia, came to New York City in 1900 at the age of nine. When his father died in 1906, David found work as an office boy at the American Marconi Wireless Telegraph Co. He quickly worked his way up, learning to become a skilled telegraph operator and becoming chief inspector in 1913. When the Radio Corporation of America (RCA) was formed in 1919, absorbing the patents of General Electric and American Marconi, Sarnoff became commercial manager of RCA. In 1930, just shy of his thirty-ninth birthday, David Sarnoff became president of RCA.

As early as 1915, Sarnoff had proposed selling radio music boxes and broadcasting music into people's homes. World War I diverted radio activity to military uses, but by 1921 Sarnoff was anxious to demonstrate the possibilities of radio broadcasting. He chose the heavyweight boxing match between Jack Dempsey and Georges Carpentier, being fought in Jersey City, New Jersey on July 2, 1921, as a big attention-getter. He set up radio receivers and loudspeakers in about a hundred theaters and social clubs up and down the east coast, and during the fight, an announcer at ringside gave a blow-by-blow account of the fight into a telephone connected to a transmitting station not far away. An engineer at the station repeated the reports into a microphone for broadcast. The episode was a huge success, with over 300,000 hearing the broadcast.

Over the next two years, radio broadcasting exploded. In 1922, 100,000 radio receivers were sold, and there were 30 radio stations. In 1923, 500,000 radio receivers were sold, and there were 556 radio stations. Radio had arrived, but who was the father? Clearly, not one person. Just as with patents, no one person really invents anything. There are always dozens, maybe hundreds, of other people's ideas and inventions that contribute to that little inventive spark, a spark that should be nourished with more hard work developing it, rather than being extinguished by the waste of time and money chasing the patent rabbit down the hole.

The radio patents marked a major change in the way inventors worked. De Forest, Fessenden and Armstrong all eventually sold their patent rights to various wireless companies. When RCA was formed, Sarnoff's goal was to acquire all of the radio patents from these companies in exchange for equity in RCA. Companies making radios would have to pay RCA a license fee, providing RCA with a continuous stream of income. RCA set up its own research laboratory, hiring the best scientists and engineers, who worked for a salary, signing over all patent rights to RCA.

This is the way it largely works today. Engineers join a company, and their job is to invent things. All patent rights belong to the company—the engineer inventor getting a certificate of recognition, or in the rare case, a bonus. But engineers are paid well, the good ones very well.

Now it is the companies that own the patents and have to hire expensive legal teams to fight the court cases and trade and cross-license patents with other companies—a huge waste of time, money and effort for no real benefit. Had all of the so-called inventions been made public through normal publication channels, the inventor would get more recognition, plus access to new beneficial ideas.

The suffocating effect of the patent system was well illustrated during World War I, when the government stopped all patent litigation and let everyone use whatever invention they liked, so as to accelerate the development of radio for use in the war. The result, as expected, was much more rapid development than before the war. Of course, after the war, the patent system was reinstated, resulting in more wasted time and money in court, with a corresponding slowdown in creative activity. One would think that it shouldn't take a hundred years to realize that the patent system should be abolished.

If David Sarnoff wasn't the *Father of Radio*, in 1950 he got the Radio and Television Manufacturers Association to grant him the title *Father of Television*. Just as RCA had acquired all of the radio patents, Sarnoff tried to get all of the television patents for RCA. The single holdout for many years was the person whose bronze statue stands today in National Statuary Hall in the U.S. Capitol. Inscribed on the base of the statue are the words **Philo T. Farnsworth**, *Father of Television*. Farnsworth conceived the idea of a scanning electronic television system while plowing potato fields on the family ranch in Idaho when he was fourteen years old in 1920. Over the next seven years, he worked on the design and in 1927 got a crude model working in his San Francisco laboratory. He applied for a patent on January 7, 1927, and two patents, number 1,773,980 and number 1,773,981, for a *Television System* and a *Television Receiving System* were awarded to Farnsworth on August 26, 1930. These two patents withstood numerous court cases, and in 1939, RCA agreed to pay royalties to Farnsworth Television and Radio Corporation—the first time RCA paid royalties rather than collected them. But both companies needed technology from the other, so after years of court litigation, a deal had to be struck.

Patent litigations are not only time-consuming and expensive, they often lead to the wrong result. The bitter fight between Lee de Forest and Edwin Howard Armstrong over the rights to Armstrong's regenerative receiver patent went on for nearly twenty years, with Armstrong winning on the merits in lower courts, only to lose to de Forest on legal technicalities in higher courts. When it went to the Supreme Court in 1928, the court never considered the merits of the case and ruled in favor of de Forest, referring to the same legal case that had been misapplied in the appeals court. The case returned to the Supreme Court in 1934 through a different route, and this time Justice Cardozo tried to understand the science behind the invention. He based his decision in favor of de Forest on a complete misunderstanding of the science —a complete blunder. Every technical expert and professional society—even the New York Times—knew that Armstrong was the real inventor, and that a great legal injustice had occurred.

Armstrong had made a lot of money from his patents, but by 1954, years of litigation had made him bitter and had drained much of his wealth. Following a fight with his wife, he committed suicide by jumping from his thirteenth-floor bedroom window. His widow, Marion Armstrong, continued the litigation of his FM patents, suing numerous companies and winning large monetary settlements, allowing her to live the rest of her life in luxury. Her final victory came from the Supreme Court on October 9, 1967, fifty-three years after Edwin Howard Armstrong obtained his first patent.

Chapter 9

Who Invented the Computer?

> *3.1.2 Eckert and Mauchly did not themselves invent the automatic electronic computer, but instead derived that subject matter from one Dr. John Vincent Atanasoff.*
>
> U.S. District Judge Earl R. Larson
> *Findings of Fact, Conclusions of Law,*
> *And Order for Judgement*
> October 19, 1973

During World War II, the Army used teams of women, called human computers, to carry out the long calculations required to create artillery firing tables. The Army was desperate to speed up these calculations and contracted with the Moore School of Electrical Engineering at the University of Pennsylvania to develop an electronic method for doing these calculations. A group headed by **J. Presper Eckert** and **John W. Mauchly** designed and built an Electronic Numerical Integrator and Calculator (ENIAC). In 1944, under contract from the Ordnance Department, they began work on a follow-on computer, called the Electronic Discrete Variable Computer (EDVAC), which was to be the first stored program computer. However, disagreements over patent rights caused Eckert and Mauchly to resign from the Moore School in 1946 and to start their own company, the Electronic Control Company, with the goal of producing a Universal Automatic Computer, the UNIVAC. Financial problems caused them to reorganize as the Eckert-Mauchly Computer Corporation in 1948 and to finally sell out to Remington Rand in 1950. The first UNIVAC was delivered to the Census Bureau in 1951.

Eckert and Mauchly had filed for a patent on the ENIAC on June 26, 1947. Seventeen years later, on February 4, 1964 the Patent Office finally granted them Patent No. 3,120,606, assigned to Sperry Rand Corporation, Sperry Corporation having acquired Remington Rand in 1955. One reason for the long delay may have been that the patent is 207 pages long, with 91 pages of drawings and 148 claims. It even has a table of contents, detailing its eleven chapters.

In 1967, Sperry Rand and Honeywell sued each other, Sperry Rand charging Honeywell with patent infringement, and Honeywell claiming the ENIAC patent to be invalid. The trial extended over nine months in 1971-1972 and created a transcript of over 20,000 pages, costing the two companies combined over eight million dollars. Finally, on October 19, 1973, Judge Earl R. Larson issued a 248-page decision. The upshot was that all claims in the ENIAC patent were found to be invalid for a variety reasons, the most controversial being that Mauchly learned critical information on how to make the ENIAC from **Dr. John Vincent Atanasoff**.

John Atanasoff was a professor of mathematics and physics at Iowa State College, who in 1938, had conceived the design of an automatic digital computer that would solve a large system of linear algebraic equations. In 1939, he obtained a grant of $650 from Iowa State College, allowing him to hire an electrical engineering graduate student, **Clifford Berry**, to help construct the computer. In December 1939, they demonstrated a working prototype, which led to a follow-on grant of $5000 to construct a full-scale computer. This full-scale computer, about the size of a desk, was completed by the end of 1941 and worked perfectly, except for some problems with the output card-writing system. Both Atanasoff and Berry had to leave Iowa State in 1942, due to the United States entry into World War II. After the war, Atanasoff never returned to work at Iowa State, his computer was dismantled and never patented. He went on to work on other things.

Atanasoff's computer was largely forgotten, mentioned in some newspaper articles in the 1940s and in a 1966 book on *Electronic Digital Systems* by R. K. Richards. This may have been where the Honeywell lawyers learned about Atanasoff, and they asked him to testify at the trial against Sperry Rand. Atanasoff proved to be an effective witness and was able to prove in court that John Mauchly had visited his lab at Iowa state in 1941, stayed in his home for five nights, observed the operation of Atanasoff's computer, and read the 35-page manuscript that Atanasoff had written, detailing the construction and operation of the computer.

Judge Earl Larson's 1973 decision in the trial of Honeywell, Inc. vs. Sperry Rand Corp. meant that no one owned a patent on the computer—it was in the public domain—probably where it belonged all along.

The idea of using a machine to calculate goes back a long way. In 1822, **Charles Babbage** built a working model of his *difference engine*, a mechanical device that could compute 6-digit numbers in a mathematical table using a *method of differences*. This was a special-purpose computer, just like Atanasoff's, designed to perform only one type of calculation. In 1834, Babbage conceived of a more powerful *analytical engine* that could be programmed to solve any mathematical problem. He spent the rest of this life designing this analytical engine, producing thousands of pages of notes, but never could build it due to lack of funds. His mechanical design included many of the components of present-day computers, including a store (memory) and a mill (Central Processing Unit—CPU). The design of a working electronic computer would have to wait a hundred years.

But even Atanasoff's computer wasn't the first. In 1938, **Konrad Zuse**, a German engineer, constructed the Z1, the first binary calculating machine, and in

1941, he completed the Z3, a general-purpose electromechanical calculating machine.

As with other inventions, the idea of computing was *in the air*, and scientists were publishing papers about it. In 1936, **Alan M. Turing**, an English logician, published a paper, *On Computable Numbers*, which demonstrated that arbitrary computations can be made with a finite state machine. Turing played a significant role in the development of early computers in England during and after World War II.

In 1937, **George Stibitz**, a physicist at Bell Telephone Laboratory, built binary circuits using relays that could add, subtract, multiply and divide. In 1938, **Claude Shannon**, based on his master's thesis at MIT, published *A Symbolic Analysis of Relay and Switching Circuits*, in which he showed how symbolic logic and binary mathematics could be applied to relay circuits.

All of these ideas, and many more, go into the design and construction of a digital computer. Mauchly undoubtedly picked up some ideas from seeing Atanosoff's computer and talking with him. But Atanosoff's computer was the size of a desk, while the ENIAC was the size of a room. Atanosoff's computer could do only one thing—solve a system of linear equations—while the ENIAC could be reprogrammed by changing the hardware configuration. Atanosoff's computer was forgotten and played no role in the design of subsequent computers, while the ENIAC continued to be used to solve a variety of different computational problems and influenced the design of follow-on computers like the UNIVAC.

Inasmuch as no one ended up with a patent on the computer after thirty years, was it helpful to have a patent system? Could the time and money wasted on useless litigation have better been spent developing smaller, cheaper and more powerful computers?

Chapter 10

Squeezing a Computer into a Smartphone

> *At present [a minimum cost] is reached when 50 components are used per circuit... The complexity for minimum component cost has increased at a rate of roughly a factor of two per year.... there is no reason to believe [this rate] will not remain nearly constant for at least 10 years. That means by 1975, the number of components per integrated circuit for minimum cost will be 65,000.*
>
> Gordon E. Moore
> *Electronics*, Vol. 38, No. 8
> April 19, 1965

The ENIAC was 100 feet long and contained over 17,000 vacuum tubes. Inside your smartphone is a computer that contains over two billion transistors. How did this happen?

On December 24, 1947, **William H. Brattain** drew a sketch of a solid-state circuit in his lab notebook and wrote, "This circuit was actually spoken over and by switching the device in and out a distinct gain in speech level could be heard and seen on the scope presentation with no noticeable change in quality... the power gain was the order of a factor of 18 or greater. Various people witnessed this test... the demonstration occurred on the afternoon of Dec. 23, 1947." Brattain had obtained voltage and power gain for the first time in a similar circuit on December 16, 1947, marking the birth of the point-contact transistor. This discovery of solid-state amplification would change electronics forever, signaling the beginning of the end of vacuum tubes as active elements in electronic circuits.

Following the end of World War II, a research team at Bell Laboratories set out to study the behavior of semiconductors, with the goal of developing new and improved components for communication systems. In addition to Brattain, other members of the team included **John Bardeen** and **William Shockley**. It had been known for a long time that a metal wire in contact with a semiconducting material could rectify electrical signals, i.e. could act as a diode. This phenomenon was discovered by **Karl Braun** in 1874. Braun would go on to build the first cathode ray tube (CRT) in 1897, invent the crystal diode rectifier (called a cat's whisker diode) in 1898, and for his contributions to the development of wireless telegraphy, share the Nobel Prize with Marconi in 1909. Crystal rectifiers were studied extensively during World War II because of their use in radar systems. The development of quantum mechanics during the first half of the twentieth century provided new theoretical insight into how semiconductors worked. Thus, the time was ripe at the end of World War II for a research program in semiconductors to be started at Bell Laboratories. Brattain, Bardeen and Shockley shared the Nobel Prize in Physics in 1956 for the invention of the transistor.

In 1949, the U.S. Justice Department filed an antitrust lawsuit against American Telephone and Telegraph (AT&T), the parent company of Bell Laboratories, alleging unfair advantage of their near monopoly in telecommunications and the reselling of phones manufactured by their subsidiary Western Electric. The suit was settled in 1956 by a consent decree, in which, among other things, all of the more than seven thousand Bell Laboratory patents were to be licensed free of charge to all U.S. companies, including patents on the transistor, solar cell, and cellular phone technology.

This freeing up of the transistor patents had the beneficial effect of accelerating the adoption of transistor technology, especially by small start-up companies—just as one would expect if there were no patent system. Sensitive to the ongoing antitrust suit, Bell Labs had already licensed the transistor patents, at relatively low costs, to any company, and provided symposia for licensees to learn the technology. In 1956, William Shockley formed Shockley Semiconductor in Palo Alto, California to manufacture silicon transistors. Unable to work under Shockley's management style, eight of his best scientists, including **Gordon Moore** and **Robert Noyce**, quit and formed Fairchild Semiconductor as a division of Fairchild Camera and Instrument. In 1968, Gordon Moore and Robert Noyce founded Intel Corporation. In 2001, Gordon Moore, commenting of the importance of the 1949 antitrust suit, stated,

> One of the most important developments for the commercial semiconductor industry... was the antitrust suit filed against Western Electric in 1949... which allowed the merchant semiconductor industry "to really get started" in the United States... there is a direct connection between the liberal licensing policies of Bell Labs and people such as Gordon Teal leaving Bell Labs to start Texas Instruments and William Shockley doing the same thing to start, with the support of Beckman Instruments, Shockley Semiconductor in Palo Alto. This... started the growth of Silicon Valley.

Jack S. Kilby was an electrical engineer who joined Texas Instruments in 1958, after learning to build transistors at Centralab. He quickly realized that trying to build microminiaturized circuits by connecting many tiny transistors, resistors and capacitors together posed serious problems. His proposed solution was to fabricate transistors, resistors and capacitors together with their interconnections on a single semiconductor device—an integrated circuit. He filed the patent for *Miniaturized Electronic Circuits*—the first integrated circuit—on February 6, 1959. It was issued as patent number 3,138,743 on June 23, 1964.

Meanwhile, Robert Noyce, while still at Fairchild Semiconductor, filed for a patent entitled *Semiconductor Device-and-Lead Structures* on July 30, 1959. The patent, which dealt with ways to more easily fabricate multiple small semiconductor devices on a single body of material, was issued on April 25, 1961 as patent number 2,981,877. It is interesting to note that while the Noyce patent was filed over five months after the Kilby patent, it was issued over three years before the Kilby patent. In 1962, Texas Instruments declared an interference against Noyce's patent claims. The patent examiner and the Board of Patent Interferences ruled in Kilby's favor. However, in November 1969, the Court of Customs & Patent Appeals overturned this ruling and upheld Noyce's claims, based, in part, on such legal minutia as whether Kilby's term "laid down" was equivalent to Noyce's term "adherent to." Does it make any sense to say that one person, who filed a patent a month before another, is somehow the true inventor of something as broad as integrated circuits, which was 'in the air" at the time? As Jack Kilby himself wrote in his July 1976 article in the IEEE Transactions on Electron Devices entitled, *Invention of the Integrated Circuit*,

> "This progress is not the work of any single individual or small group of individuals. It has come about because of the contributions of thousands of engineers and scientists in laboratories and production facilities all over the world."

This was after Gordon Moore's prediction of 65,000 components per integrated circuit by 1975 had come true. Moore modified his prediction in 1975, now called *Moore's Law*, to say that instead of doubling every year, the number of components on an integrated circuit would continue to double every eighteen months. It turns out that Moore's Law has pretty much remained true ever since. That's why you now have a computer inside your smartphone containing over two billion transistors.

It was this ability to add more and more transistors on a single semiconductor chip, year after year, at an exponential rate of growth, that fueled the computer revolution and the huge economic growth of Silicon Valley. It is what allowed **Marcian ("Ted") Hoff** to design Intel's first microprocessor in 1971—a true computer on a chip. It is what allowed **Steve Wozniak** and **Steve Jobs** to introduce the Apple II computer in 1977, and IBM to introduce its PC in 1981. It is what allowed **Bill Gates** to start

Microsoft to produce software to run on these new personal computers—personal computers that grew smaller and more powerful each year, leading to portable computers, then laptop computers, then iPads, and finally your smartphone.

As Jack Kilby wrote back in 1976, all of this remarkable development required the inventiveness of thousands of engineers and scientists. Many of these engineers will have their names on patents assigned to their companies, patents that are often obsolete by the time they are issued, as the power of Moore's Law pushes the industry ahead at an accelerated pace.

Is it any wonder that courts have trouble determining the validity of a patent, when the ideas of thousands of engineers and scientists have made the invention possible?

Chapter 11

Instant Photography Ideas

> *Effective January 9, 1986, Kodak, its officers, agents, servants, employees, and attorneys, and those persons in active concert or participation with them who receive actual notice of this judgment by personal service or otherwise, are enjoined and restrained from infringing any one or more of the following ... [Polaroid patents] including, without limitation, by manufacture, use or sale of PR-10 film and EK-4 and EK-6 cameras.*
>
> > Judge Rya Zobel
> > Official Judgment
> > Polaroid Corp. vs. Eastman Kodak Co.
> > October 11, 1985

It was Tuesday, October 6, 1981, and **Edwin H. Land**, the 72-year-old founder of Polaroid Corporation, was in his second day on the witness stand under direct examination by Polaroid's lead trial counsel **William K. Kerr**. Polaroid had filed a patent infringement complaint against Kodak over five years earlier, on April 26, 1976. Dozens of Polaroid and Kodak engineers and scientists had been deposed during this long five-year pre-trial discovery period. The trial judge, **Rya Zobel**, was the fourth judge assigned to the case, taking over on March 17, 1980. She would be the sole person to decide this case, and on Land's second day of testimony, the judge seemed interested in learning as much about the technical details of instant color-film photography as she could; and Land was more than happy to explain to her, in a patient manner she seemed to appreciate, the special features of his patents.

Land's direct testimony would go on for a third day, and then Kodak's lead trial counsel **Frank Carr** would cross-examine Land for ten days, testing the patience of Judge Zobel. The 74-day trial ended on February 25, 1982. Judge Zobel deliberated for three and a half years, before issuing her opinion on September 13, 1985 in favor of Polaroid. She ruled that eight of the ten Polaroid patents were valid and that Kodak had infringed seven of them. The following month, on October 11, 1985, she issued a permanent injunction, ordering Kodak to cease selling their instant cameras, of which they had already sold sixteen million. Shockingly, she refused to stay the injunction pending appeal.

Kodak scrambled to appeal the stay before it took effect on January 9, 1986. The appeals court waited until almost the last minute, and on January 7, 1986 refused to vacate the injunction or the stay. A final, desperate appeal to the U.S. Supreme Court was denied on January 8th. The decision of a single judge had put Kodak out of the instant photography business, with the resulting loss of thousands of jobs and huge customer disruption.

But Kodak's problems weren't over yet—the damages phase of the litigation was about to begin. After two years of discovery, Judge Zobel was replaced with **Judge A. David Mazzone**, who presided over the trial that began on May 1, 1989 and went on for six months, ending on November 20, 1989. Now it was Judge Mazzone's turn to deliberate for almost a year, before announcing his decision on October 12, 1990, which ordered Kodak to pay Polaroid almost a billion dollars in damages. The two sides finally settled the case for $925 million on July 16, 1991—fifteen years, three months and twenty days after Polaroid filed its complaint in the U.S. District Court in Boston. Edwin Land had died over four months earlier on March 1, 1991.

Did the patent system work in this case? First of all, a system that takes over fifteen years to settle a dispute is a broken system almost by definition. The Polaroid employees, many of whom shared in the $925 million windfall, may have thought the system worked, until many of them began receiving pink slips, as Polaroid slid into bankruptcy a decade later—overtaken by digital photography. Of course, Kodak employees thought the patent system failed in this case. But Kodak, too, was overtaken by the digital photography revolution, even though Kodak's research labs had developed some of the first digital cameras. Kodak was too invested in film, and filed for bankruptcy in 2012.

Edwin Land always strongly supported the patent system, believing it was essential to Polaroid's ability to succeed in the market place. But was this true? Could Land have developed his first Polaroid camera without patent protection? Almost certainly yes. **Ronald K. Fierstein's** book, *A Triumph of Genius—Edwin Land, Polaroid, and the Kodak Patent War* is a detailed account of Polaroid's history and its fifteen-year patent war with Kodak. It is clear from this book that developing the various Polaroid cameras was difficult, requiring enormous effort, not something that a competitor could easily duplicate, as Kodak, even with all of its expertise and knowing the details of Polaroid's patents, learned when it tried to produce a camera that would compete with Polaroid's SX-70. But Polaroid could not have developed its first instant camera without the help of Kodak, who developed and produced a special negative for the camera.

For many years, Polaroid had little real competition in the instant photography business. This may have been its Achilles' heel. Competition is actually good. It forces a company to focus on the customer and to keep up with technological developments.

In 1975, a year before Polaroid sued Kodak, **Steven Sasson** was working on an *Electronic Still Camera* at Kodak. The camera used a Fairchild charged-coupled device (CCD) that captured an image on a 100 x 100 array of pixels and stored the image on a magnetic tape cassette for replay on a TV screen. Kodak filed for a patent on this device on May 20, 1977, and patent number 4,131,919 was issued on December 26, 1978. **Willis Adcock** at Texas Instruments had filed for a patent on a similar *Electronic Photography System* on October 29, 1976, and that patent, number 4,057,830, was issued on November 8, 1977.

Kodak continued working on digital camera systems (DCS), and in 1991, the year its legal battle with Polaroid finally came to an end, it introduced the DCS 420 digital camera for the professional photographer, which had a 1.3-megapixel sensor and a Nikon F3 body. But Kodak couldn't compete with companies such as Sony, Casio, Olympus, Nikon, and Cannon in providing digital cameras for the general public. The first camera that was built into a phone was the Kyocera VP-210 in 1999, which could hold twenty still shots. Sales of digital cameras began slowly around 2001, peaked about 2010, and then dropped precipitously as the sales of smartphones, introduced around 2004, skyrocketed. The sales of film cameras were basically dead by 2008 and by mid-2016 when Apple sold its one billionth iPhone, regular digital cameras were on life-support.

So perhaps neither Kodak nor Polaroid could have survived, unless they had gotten into the smartphone business. Incidentally, in 1992, seven years after **Steve Jobs** had quit Apple in a struggle with its president, **John Sculley**, Apple was in deep financial trouble. Sculley's advice to the board was to sell Apple to a larger company such as Kodak or AT&T. That obviously never happened, Sculley was out the following year, Jobs returned to Apple as Interim CEO in 1997, the iPod was introduced in 2001, and the iPhone in 2007.

Chapter 12

What Does a Smartphone Look Like?

> *If I'm the juror, I just don't know what to do. I'd have the iPhone in the jury room; I'd—look at it. I just wouldn't know.*
>
> > Justice Anthony Kennedy
> > Oral Arguments, Supreme Court
> > Samsung Electronics vs. Apple, Inc.
> > October 11, 2016

It was 10:05 a.m. on Tuesday, October 11, 2016, and **Chief Justice John Roberts** spoke from the bench: "We'll hear argument first this morning in Case No. 15-777, Samsung Electronics v. Apple, Incorporated. Ms. Sullivan."

Kathleen M. Sullivan, the lawyer arguing on behalf of the petitioners, Samsung, addressed the court:

> Mr. Chief Justice, and may it please the Court: A smartphone is smart because it contains hundreds of thousands of the technologies that make it work. But the Federal Circuit held that Section 289 of the Patent Act entitles the holder of a single design patent on a portion of the appearance of the phone to total profit on the entire phone.
>
> That result makes no sense. A single design patent on the portion of the appearance of a phone should not entitle the design-patent holder to all the profit on the entire phone.

For the first time in 120 years, the Supreme Court was hearing a dispute over a design patent. A design patent, unlike the more common utility patent, covers only "the visual ornamental characteristics embodied in, or applied to, an article of manufacture." In April 2011, Apple had sued Samsung for infringing several of its utility and design patents, and Samsung countersued, accusing Apple of infringing some of its patents. A jury trial was held in 2012, and on August 24, 2012 the jury returned a verdict, awarding Apple over a billion dollars in damages. After appeals, corrections of miscalculations, and a retrial of damages, the amount due Apple was considerably reduced, and by 2015 only $399 million was in dispute, related to Apple's design patent. The calculation for damages was based on total profits from Samsung's smartphone sales, and Samsung argued that it should be based only on the design, that is, the look of the iPhone case and not everything that was inside the smartphone. This is what the U.S. Supreme Court would spend an hour talking about on the morning of October 11, 2016.

By this time, the justices of the Supreme Court had reviewed the briefs, including a "Brief *Amici Curiae* of 50 Intellectual Property Professors in Support of Petitioners," which stated, in part,

> Nor does all, or even most, of the value of a product normally come from patented designs. People don't buy iPhones for their appearance alone; they buy them for their functions. Those functions contribute substantially to the phone's value and they are covered by many utility patents. Indeed, by one estimate, there are 250,000 patents that arguably cover various aspects of a smartphone. To conclude that one design patent drives the purchase of the product, and therefore that the defendant's entire profit is attributable to infringing that patent, is to say that none of those functional features contribute anything to the value of the phone—a ludicrous proposition.

To get a sense of the deep legal issues confronting the court that morning, the Volkswagen Beetle makes it into the following exchange between **Justice Anthony Kennedy** and Ms. Sullivan:

> JUSTICE KENNEDY: So we find out the—the production cost if—if a billion dollars were spent on the inner parts and a hundred million was spent on the face, then it's a 10:1 ratio.
>
> MS. SULLIVAN: That's absolutely right, Your Honor. Apple—
>
> JUSTICE KENNEDY: So you'd have expert testimony on all of that.
>
> MS. SULLIVAN: Yes, Your Honor, you would. And you would—but that's just one way
>
> JUSTICE KENNEDY: Suppose—suppose you had a case where it's a stroke of genius, the design. In—in two days, they come up with a design—let's let's assume the Volkswagen Beetle analogy that some of the briefs refer to. Suppose the Volkswagen Beetle design was done in three days, and it was a stroke of genius and it identified the car. Then it seems to me that that's quite unfair to say, well, we give three days' profit, but then it took 100,000 hours to develop the motor.
>
> MS. SULLIVAN: Well, Your Honor, here's what we would do with the Beetle.
>
> JUSTICE KENNEDY: I mean, that's what—it seems to me that that's what you would be arguing.
>
> MS. SULLIVAN: It's not, Your Honor. To answer Justice Ginsburg's question, there are three ways Apple could have but did not even attempt to prove the total profit from the relevant article of manufactures here, the front face, or the display screen. One could have been accounting. One could have been consumer demand evidence, Justice Kennedy, as

you suggested. Apple could have said well, people really like the front face disproportionately to all the other parts of the phone, so they could have used consumer survey evidence to prove that. But—and so accounting evidence or indirect evidence through consumer survey. But, Your Honor, as to the Beetle, we concede that the total profit from the article of manufacture may sometimes be a substantial part of the total profit on the product.

Let's take the Beetle, or let's take a cool, shark-shaped exterior body on a car like the Corvette. It may be that the article of manufacture to which the design patent is applied is just the exterior body of the car, but it may be that nobody really wants to pay much for the innards of the Corvette or the Beetle. They want to pay for the cool way it looks.

If that's so, it should be open to the patent-holder to prove that the bulk of the profits come from the exterior of the car.

Justice Samuel Alito then got in a question.

JUSTICE ALITO: Is there any difference in practical terms between that and your causation argument or apportionment?

MS. SULLIVAN: Yes, Justice Alito.

JUSTICE ALITO: What is the difference?

MS. SULLIVAN: The difference is we concede under article of manufacture that the holder of the patent gets profit from the article, even if the profit does not come entirely from the design.

Let me give you an example with a phone's front face. Consumers may value the front face because it's scratch-resistant, because it's water-resistant, because it's shatterproof. We're going to give the patent-holder under our article-of-manufacture test all the profits for the front face, even if it includes profit from those non-design features of the front face, where the pure apportionment test or pure causation test would limit the profits to the profits from the design parts rather than the functional parts. So, Your Honor, that's a little bit overinclusive. We're getting a little more with article of manufacture than we do with a pure causation test, and plaintiffs should be happy for that.

But the reason we think it's consistent with Congress's purpose, Your Honor, is that what Congress was trying to do was provide a rule that gives design-patent holders total profit from the article of manufacture.

Later on, **Justice Ruth Bader Ginsburg** had a question.

JUSTICE GINSBURG: Did Samsung, at the trial, propose basing damages on profits from an article less than the whole phone?

MS. SULLIVAN: Six times, Your Honor. And we were rebuffed every time. At the—in the jury instruction—sorry. At the—before the trial began, we submitted a legal brief. It's Docket 1322. We said very clearly article of manufacture is less than the total phone and profit should be limited to the profit from the article. We said again in the jury instructions—and here I would refer you respectfully to joint Appendix

206, 207 and to the result of that on petition Appendix 165A. What happened is we went to the court and we said please listen to us about article of manufacture, if you only get the total profit on the article. The district court said, no, I already said no apportionment back in the Daubert. Because I said no apportionment, she shut us out of both theories. The district court shut us out of article of manufacture as the basis for total profit, and it shut us out of causation or apportionment, which we don't press here.

So that's twice. Our legal brief, our charge conference. And then again in our 50A and the key rulings on 50A at the close of evidence, we again said article is separate from apportionment, and the article here is less than the phone. At 197 we said at—sorry. At JA197 we again said article is less than the phone. And in the 50B at the close of the first trial, we again said article is less than the phone.

Second trial happens on certain phones. Again, in the 50A and the 50B, the trial court says again, I have ruled that there's no apportionment for design patents. You cannot talk to me about article of manufacture. We tried over and over and over again to get the article of manufacture's theory embraced, and we were rejected. And why does that matter, Your Honor? Because there was evidence in the case from which a reasonable, properly instructed jury could have found that the components were the front face, the front face, and the display screen. And the evidence came out of Apple's own witnesses, which we're certainly entitled to rely on. Your Honor, Apple's own witnesses again and again said what are you claiming. And when the witnesses got on to talk about infringement, they didn't say the whole phone, the look and feel. They said we're claiming a very specific front face, and by the way, ignore the home button. We're claiming a very specific front face and surrounding bezel, and by the way, ignore everything that's outside the dotted lines.

And if I could just remind you that we've reprinted the patents for you to see, and they may look like an iPhone on page 7, which is the D'677. They may look like an iPhone in the D'087, which was in Blueberry, set 8, but the claim is not for the iPhone. The claim is for the small portion of the external appearance of the phone that is inside the solid line. Apple disclaimed everything outside the solid line. It disclaimed portions of the front face with dotted lines.

And Your Honor, the question for the jury was not did people think that the look and feel of an iPhone was great. The question for the jury was did the very small portion of a smartphone that Samsung makes look substantially similar to the very small portion of the patent claim?

Now that, Your Honor, there is no basis in this record for a conclusion that the entire product, profit on the phone, corresponds to the entire profit from those articles. What Apple should have done is done either of the two things we discussed earlier, accounting evidence about revenues minus cost of goods sold on the components, or it should have done consumer survey evidence like our expert did.

After **Brian H. Fletcher**, Assistant to the Solicitor General in the Department of Justice appeared as an *amicus curiae*, supporting neither party, **Seth P. Waxman**, the lawyer arguing on behalf of the respondent Apple, got his turn. Toward the end of the allotted hour, Chief Justice John Roberts seems to have heard enough.

> CHIEF JUSTICE ROBERTS: I—maybe I'm not grasping the difficulties in the case. It seems to me that the design is applied to the exterior case of the phone. It's not applied to the—all the chips and wires, so why—
> MR. WAXMAN: That's right.

Following more back and forth, at 11:07 a.m., Chief Justice Roberts announced, "Thank you, counsel. The case is submitted."

On December 6, 2006, in a unanimous (8-0) decision, the Supreme Court reversed the judgment of the United States Court of Appeals for the Federal Circuit and remanded the case for further proceedings.

The opinion, delivered by **Justice Sonia Sotomayer**, spends over seven pages trying to define the meaning of the words "articles of manufacture," because the law specifies that "Whoever during the term of a patent for a design, without license of the owner, (1) applies the patented design, or any colorable imitation thereof, to any article of manufacture for the purpose of sale, or (2) sells or exposes for sale any article of manufacture to which such design or colorable imitation has been applied shall be liable to the owner to the extent of his total profit, but not less than $250."

The court ended up agreeing with Chief Justice Roberts that, in this case, the "article of manufacture" was just the exterior case of the phone and not all the chips inside that make it useful.

On Thursday, January 12, 2017, the U.S. Court of Appeals for the Federal Circuit officially reopened the case. Finally, on June 27, 2018, Apple and Samsung informed Judge Lucy Koh that they had reached a settlement on the seven-year-long patent fight. Terms of the settlement were not disclosed.

This wasn't the only patent case that Apple has been litigating against Samsung. In addition to a second case in the United States, there have been cases around the world in South Korea, Japan, Germany, France, Italy, Australia, and Great Britain.

What is striking about this and other court cases, is that the fundamental question as to the validity of any of these patents wasn't even considered. The only question the Supreme Court considered was the meaning of the words "articles of manufacture."

The design patent in question before the Supreme Court was Apple's design patent number D593,087 for an *Electronic Device*, filed on July 30, 2007, listing fifteen inventors. The patent was issued almost two years later on May 26, 2009. As with all design patents, it contained only a single claim: "The ornamental design of an electronic device, substantially as shown and described." The patent contained 48 figures—line drawings, showing the iPhone from the top, bottom, side, end and perspective—pointing out certain elongated oval and circle shapes. There's only so much you can say about a thin rectangular box with curved corners and beveled sides. But apparently it was enough for the Supreme Court to use up an hour of oral argument, determining whether Samsung's similar thin rectangular box design should require them to pay Apple $399 million. The Supreme Court said, "Not that much," but the case dragged on for almost two more years.

It should be obvious that the shape of the thin rectangular box is not what makes the smartphone smart. What does?

Chapter 13

What Makes a Smartphone Smart?

> *We chose our system name, Google, because it is a common spelling of googol, or 10^{100} and fits well with our goal of building very large-scale search engines.*
>
> Sergey Brin and Lawrence Page
> *The Anatomy of a Large-Scale Hypertextual Web Search Engine*
> Computer Networks and ISDN Systems
> Vol. 30, p. 108, 1998

What makes a smartphone smart? Two words: *hardware* and *software*. The most important hardware are the chips containing billions of transistors, resulting from Moore's law as discussed in Chapter 10. Some of the chips are special-purpose digital circuits designed to do a specific task, but there is also a powerful microprocessor that executes instructions stored in memory. These instructions are generated by writing software programs, and it is these software programs that really make the smartphone smart.

Can software programs be patented? The U.S. Patent Office started issuing patents on what could be considered to be software by the 1970s. In 1982, Congress established a new Federal Circuit to hear patent cases, and since then the number of patents issued for software has grown exponentially, with tens of thousands of such patents issued to date. This is unfortunate, as it stretches the purpose and criteria for patents to the limit, and there is considerable controversy surrounding software patents. Recall that ideas cannot be patented. In particular, abstract ideas and mathematical formulas are not patentable. The U.S. Supreme Court has struggled to reconcile these prohibitions with software patents and has struck down many such patents, while still looking for tests that will allow certain software to be patented. There are no such tests that make any sense.

A more serious objection to software patents has to do with the non-obvious requirement for a patent. Once the idea of what a software program should do has been stated—and ideas are not patentable—then any number of computer science graduate students can come up with a software program to do the job. This is how all software has been developed.

The patenting of software has led to patent trolls—unscrupulous companies that acquire patents for the purpose of suing large companies with deep pockets. In 2009, Bedrock Computer Technologies brought complaints of patent infringement against ten companies including Google, AOL, Yahoo, Amazon and PayPal. Bedrock had acquired an obscure software patent on "methods and apparatus for information storage and retrieval using a hashing technique with external chaining and on-the-fly removal of expired data." Like most software patents, this one should never have been approved by the patent office, the problem solved being the type of problem that should be given as a homework problem in a good computer science graduate course. In any event, the solution apparently ended up in some versions of the Linux kernel—Linux being an open-source operating system to which anyone can make contributions. In 2011, a jury trial for Bedrock vs. Google was held in the U.S. District Court for the Eastern District of Texas—a district known for being pro-patent and once referred to by Justice Antony Scalia in oral arguments as a "renegade jurisdiction." On April 15, 2011, the jury found that Google infringed the patent and imposed a $5,000,000 judgment. Both sides quickly settled, and on May 18, 2011, the judge vacated the verdict and dismissed both claims and counterclaims with prejudice. Does a patent system that tolerates such trolling make any sense?

The jury in this trial determined whether or not the patent was valid. In the Polaroid vs. Kodak trial, a single judge made those decisions. To be valid, a utility patent must be "non-obvious to a person having skill in the area of technology related to the invention." It is almost certain that any judge or member of a jury will not be such a skilled person. In fact, lawyers will do all they can to screen out such skilled persons from the jury pool. The way patents are written by lawyers, using an awkward format established over 200 years ago, makes them almost useless in trying to understand the invention. Once a patent attorney has translated an inventor's description into the patent legalese, the inventor often cannot understand his or her own invention.

Most patents describe "inventions" that are not that inventive or unique, and are, in fact, "obvious" to someone skilled in the art. There is no way for a patent trial to determine if an invention is obvious. The only way is to do the experiment. Remember, you cannot patent an idea. So, the idea of a smartphone is not patentable. Nor is the idea of a touchscreen, nor the idea of cellular communication, nor the idea of what you want a particular software program to do. If you take 100 bright engineers and computer scientists and tell them to implement one of these ideas, they will do it. That's what they do. That's their job—what they get paid for. Some of the solutions to the problem may be the same; some will be different. Some solutions may be cleverer than others, but there won't be any solution that is "non-obvious" to the bright engineers who are "skilled in the art." If this is the case, then why is any patent valid?

There are thousands of software programs in your smartphone, which make it smart. One that you may use a lot is a search engine.

Larry Page and **Sergey Brin** were computer science Ph.D. students at Stanford University in 1996 looking for a research topic. They developed a search engine that they called Google with the goal of being able to search the entire world wide web. The search engine used a web crawler to collect the words from web pages and save them in a sophisticated index. Each web page was ranked using a system they called PageRank. The highest ranked pages would be the ones that showed up at the top of the search list. In 1998, Larry Page and Sergey Brin left Sanford and started their company Google.

On January 8, 1998, Larry Page had filed for a patent on his PageRank system entitled, *Method for Node Ranking in a Linked Database*—essentially a software patent. The patent, number 6,285,999, which was assigned to Stanford University, was issued on September 4, 2001. Stanford had granted Google an exclusive license to use this patent in return for 1.8 million shares of Google stock, which Stanford sold in 2005 for $336 million.

This was certainly a windfall for Stanford University, and you might think it blows a hole in my claim that the patent system should be abolished. Not at all—it is actually a big reason why the patent system *should* be abolished!

First of all, the idea of graduate students—and faculty members, for that matter—seeking patents is antithetical to open enquiry, which is the hallmark of higher education. The purpose of a patent is to *exclude* someone from using a particular idea; the purpose of higher education is to spread ideas far and wide. When graduate students or faculty members hide ideas from other graduate students or faculty members for fear of someone else disclosing or highjacking the idea before they have a chance to file for a patent, higher education has lost its way. The new patent law of 2011 that changed "first to invent" to "first inventor to file" makes things even worse by forcing inventors to hide their inventions from others, because if others disclose it before the inventor files, the patent is invalid.

Secondly, the research of Page and Brin at Stanford was supported by a grant from the National Science Foundation. We, as taxpayers, had as much right to the results of their research as Stanford University. Where was our cut of the $336 million?

Finally, as with all software patents, the validity of the patent needs to be questioned. The PageRank algorithm is a good idea and produced results arguably better than the several search engines already in use at the time. But Larry Page didn't get the idea out of the blue, all by himself, while plowing a potato field at the age of fourteen, like Philo Farnsworth and television. He certainly talked to Sergey Brin and his thesis advisor and read the twenty-odd technical papers and seven related patents that he references in his patent. And all of those technical papers had references to previous work, containing ideas upon which future work was built. Of course, all the graduate computer science courses he took, plus all the undergraduate computer engineering courses he took at the University of Michigan prepared him to be able to conceive of an algorithm to solve his particular problem. Without all of the research papers and textbooks that he had read, there would be no PageRank algorithm, no patent, and no Google. When he obtained good research results, his obligation was to publish these results, just as others had done from which he benefitted.

Does this mean that without a patent and a patent system he couldn't have started Google? Of course not. His Google search engine gave better results than the other competitive search engines, and so his was likely to succeed. Of course, he would have to continually improve it, as Google has done, the current ranking algorithm most likely being quite different from the one in his PageRank patent.

But, what about Stanford? Do they lose their $336 million? Remember, they didn't get the money directly from the patent. They got the money by selling Google stock. They could have invested some of their $22 billion endowment in Google stock, or Page and Brin, grateful for the education they had received at Stanford, could have donated Google stock to Stanford's endowment fund. This is how universities traditionally grow their endowment funds—from grateful alumnae.

Chapter 14

Storing Lots of Photos on Your Smartphone

> *For digital image applications involving storage or transmission to become widespread in today's marketplace, a standard image compression method is needed to enable interoperability of equipment from different manufacturers.*
>
> Gregory K. Wallace
> *The JPEG Still Picture Compression Standard*
> IEEE Transactions on Consumer Electronics
> Vol. 38, February 1992

Robert M. Fano was teaching an electrical engineering graduate course on information theory at MIT in 1951. He told the students that they could either take the final exam or write a term paper on finding the most efficient way to encode a message containing numbers, letters and other symbols using binary codes—sequences of ones and zeros. Fano had worked on this problem, along with his colleague **Claude E. Shannon**, but he knew that his method was not optimal, so why not give his graduate students a chance to solve the problem.

This is a technique I've also used in my teaching. Years ago, while teaching a graduate course in coherent optics, I gave as a take-home exam a research problem I had been working on. Sure enough, one of the bright graduate students came up with a better solution than mine.

The graduate student in Fano's class who came up with a better solution was **David A. Huffman**. The story is told in a profile of Huffman that appeared in the September 1991 issue of *Scientific American*. After thinking about the problem all term and about to give up, reconciled to taking the final exam, Huffman suddenly saw the solution. He knew that the basic idea was to have fewer bits (ones and zeros) represent the more commonly used characters and to use longer sequences of bits to represent those characters occurring less frequently. He also knew that the coding method could be represented as a binary tree in which the code for each character would be given by following a particular path down the tree—taking the left path at a node would add a 0 to the code, and taking the right path would add a 1. Each character was a leaf on the tree, and the further down the tree you went, the longer would be the code. Therefore, the characters used most often should have shorter paths to the leaf, while rarely used characters should be at the bottom of the tree, with the longest paths. The question was how to build the tree?

Fano's solution to the problem, known as Shannon-Fano coding, used a top-down method, starting with the highest probability characters. The trick that Huffman recognized was to build the tree bottom up, starting with the lowest probability characters. When Huffman showed Fano his solution, Fano reportedly said, "Is that all there is to it?"

Huffman coding is widely used today for compressing digital data in many different applications. We will see that it is used in the third step of the JPEG compression algorithm, used for reducing the size of digital images.

Huffman published his coding method in the September 1952 issue of the *Proceedings of the I.R.E.* (Institute of Radio Engineers). He never tried to patent any of his ideas, although years later, many patents would be issued on various compression algorithms. Huffman went on to become a professor at MIT and later headed a new computer science department at the University of California at Santa Cruz.

I recently took several photos on my smartphone. The size of the image in each photo was 3264 × 2448, which multiplies out to 7,990,272 picture elements, or pixels. The small solid-state sensor that captures the image records three color channels for each pixel: a red channel, a green channel, and a blue channel. Remember that Maxwell's first color photograph, made in 1861, used red, green, and blue color filters. A very similar technique is used in the image sensor in your smartphone to separate the light signal into three separate signals representing the red, green, and blue components of each pixel in the image. Each color signal is converted to an 8-bit digital value, called a *byte*, representing a decimal number between 0 and 255. Therefore, three bytes of data, or 24 bits, are required to store each color pixel value in an image. Therefore, each of my photos, containing 7,990,272 pixels, will require 7,990,272 × 3 = 23,970,816 bytes of memory, which is about 22.88 megabytes (MB). Note that 1 kilobyte (KB) equals 1024 bytes, and 1 megabyte (MB) equals 1024 kilobytes (KB). Also, 1 gigabyte (GB) equals 1024 megabytes (MB).

However, my images are stored as JPEG files, and the actual amount of memory used to store each photo varies from about 1.7 MB to 4.8 MB. How can this be? JPEG files are compressed files that do not store the complete original image, but the image still looks pretty good. Was any information lost? Yes, but it is generally information that your eyes cannot detect.

This means that on average I can store about ten JPEG images for every one uncompressed original image. Often the results are even better. I have a cropped 2016 × 1998 image, which would normally require 11.5 MB of memory, but the JPEG file uses only 383 KB—a 30 to 1 reduction. If you crop images and remove lots of small detail, the size of the JPEG file can go down drastically.

I have used about 25 GB of memory to store photos on my smartphone. Therefore, I could be storing about 1000 uncompressed photos, or 10,000 to 30,000 JPEG images. This is some change from when I was as teenager and had a Brownie Hawkeye camera that could take twelve pictures on one roll of black and white film! JPEG compression allows you to store thousands of photos on your smartphone. How does it work?

JPEG stands for Joint Photographic Experts Group, an international group made up of volunteers, including representatives from companies in the U.S.A., Israel, France, Germany, Japan, Denmark and the U.K. They started meeting in 1987 with the goal of defining an international standard image compression method that would enable digital photographic images to be generated and read by equipment from different manufacturers. This is necessary if users are to be able to move images from one place to another easily. Imagine posting an image on your Facebook page, which others couldn't view on their computer or smartphone. Clearly, all kinds of standards are necessary for you to enjoy all of the inventions of the modern digital age.

The patent system is an obvious roadblock to establishing meaningful standards. Government officials often have to work with standards groups to ensure that any required patent licenses are fair, adding a layer of complication and delay. Standards groups try to avoid any patent conflicts, if at all possible.

In the case of the JPEG standard, the group studied about a dozen different methods, finally selecting the one that produced the best results. The JPEG compression method involves three main steps, applied to each of the three color channels (after some possible pre-processing).

In the first step, the image is divided into 8×8 blocks of pixels, each block containing 64 bytes of data, representing a small two-dimensional square piece of the image, with eight pixels on a side. These 64 bytes of data are then transformed into 64 different bytes of data, which represent the frequency content of that image segment. This is called the *frequency domain*, and the numbers in the upper-left corner of the 8×8 frequency block represent low frequencies—meaning that the image values in the original 8×8 image block don't change much from one pixel to the next. Numbers in the lower-right corner of the 8×8 frequency block represent high frequencies—meaning that the image values in the original 8×8 image block change rapidly from one pixel to the next. In real photos, pixel values don't change that much from one pixel to the next, so low spatial frequencies are more common than high spatial frequencies. This means that the numbers in the upper-left of the 8×8 frequency block will be larger than the high-frequency numbers in the lower-right of the 8×8 frequency block.

The calculation used to transform each 8×8 image block to its corresponding 8×8 frequency block is called the *Discrete Cosine Transform,* a variation of the Fourier transform, which was introduced by **Joseph Fourier** in 1822, and is a well-known mathematical operation taught in many engineering, science and mathematics courses. Fourier transforms are widely used in many engineering applications, and there is nothing to patent here. An inverse Fourier transform operation recovers the original signal.

The second step in the JPEG compression method is to reduce the number of bits needed to represent the frequency values in each 8×8 frequency block. This is done by dividing each value in the 8×8 frequency block by some predetermined number and then rounding to the nearest integer. High frequency components that aren't important are divided by a large number, usually resulting in a final value of zero. Low frequency components, which are more important, are divided by smaller numbers. Such filtering operations are common in all types of signal processing and image processing operations. Entire courses are taught on this subject. There is nothing new to patent here.

The resulting filtered, or quantized, 8×8 frequency block will then contain a few non-zero values in the upper-left, low-frequency region of the 8×8 block. If one lists the 64 numbers in this block in a zig-zag pattern, starting at the upper-left corner, the result will be a few numbers, separated by strings of zeros, and ending with a long string of zeros. Instead of listing all 64 values, including long strings of zeros, a well-known run-length encoding method is used in which the number of repeated values is listed, rather than listing all of the values. This greatly reduces the number of bytes needed to store the data.

The final step further reduces the number of bits needed to represent the run-length encoded data by using Huffman coding, discovered back in the 1951 graduate course at MIT.

This small JPEG file then contains one long bit stream representing the filtered frequency components of the original image. This file is what is sent to a website or to your printer. To recover the image, the reverse process has to take place. The Huffman code is unscrambled to reveal the run-length coding from which all of the 8×8 frequency blocks can be recovered. The inverse Discrete Cosine Transform will then reproduce the final image.

All this seems very complicated, but it takes place millions of times a day as people all over the world share photos on various websites. Surely, no one would want to interfere with this seamless process by trying to claim patent rights on the JPEG method. Unfortunately, the patent system is such that it is easily abused in an attempt to make a quick buck.

Compression Labs, Inc. was a company founded in 1985 to produce video conferencing equipment. On October 27, 1986, they filed for a patent on

Coding System for Reducing Redundancy. The patent specifically relates to methods and apparatus used in video conferencing systems, by taking advantage of small differences between adjacent frames in a video signal. The patent, number 4,698,672, was issued on October 6, 1987. It has little relationship to the JPEG standard compression method, which deals only with still images.

Nevertheless, fifteen years later, after its hardware sales had plummeted, Compression Labs had become a subsidiary of Forgent Networks, a company set up to make money by licensing the '672 patent to anyone who was using the JPEG compression method, which, of course, by that time was everyone. Apparently, Forgent signed multi-million-dollar license agreements with some Japanese companies in 2002. In April of 2004, Forgent sued 31 companies, including Adobe Systems, Apple, Hewlett-Packard, IBM and Xerox, for infringing on their '672 patent by using JPEG files. In July of 2004, IBM, Hewlett-Packard and 20 other companies sued Forgent, seeking to invalidate the '672 patent. In 2005, the Public Patent Foundation, a non-profit organization dedicated to challenging the validity of suspect patents, asked the U.S. Patent Office to take another look at the '672 patent claims. In 2006, the patent office rejected many of the broad claims of the '672 patent, based on prior art, which the patent filers had known about, but hadn't disclosed. In November 2006, Forgent settled its patent litigation over the '672 patent for a small fraction of the $1 billion it had hoped to collect.

Do such abuses of the patent system really "promote the progress of science and the useful arts?"

Chapter 15

Using Your Smartphone as a Flashlight

Professor Akasaki, Professor Amano, Professor Nakamura:
You have been awarded the 2014 Nobel Prize in Physics for your invention of efficient blue light emitting diodes, which has enabled bright and energysaving white light sources.

>Presentation Speech by Professor Anne L'Huillier
>Member of the Royal Swedish Academy of Sciences
>Member of the Nobel Committee for Physics
>10 December 2014

Have you ever been caught using your smartphone in a restaurant? No, I don't mean being rude by texting or checking your email; I mean using your smartphone flashlight to read the menu in a dimly-lit restaurant. Lots of people find this very convenient. However, it turns out that inventing the simple little white light used in your smartphone was not an easy task, requiring decades of research by thousands of scientists and engineers, leading eventually to a Nobel Prize for three of them.

The Nobel Prize often rewards an invention of greatest benefit to mankind, and the high-efficiency, white-light light-emitting diode (LED) would qualify because of its potential to light the world far more efficiently than current light bulbs. The incandescent light bulb, invented by Thomas Edison almost 140 years ago, is only about 4% energy efficient, with most of the electrical energy converted to heat rather than light. Fluorescent tubes are more efficient than incandescent bulbs, but contain mercury, which is toxic. LEDs are far more efficient than incandescent bulbs or fluorescent tubes (over 50% efficient) and can last up to 100,000 hours, compared to 1,000 hours for incandescent bulbs and 10,000 hours for fluorescent tubes. About a quarter of the world's electrical energy consumption is used for lighting, so replacing incandescent bulbs and fluorescent tubes with LEDs can have a major impact on electrical power generation needs. For example, the low-energy requirement of LEDs means that emerging countries might be able to bypass expensive electrical grids and power LED lights directly from solar panels.

With such a potential, you would think that lots of companies would be fighting for market share and filing for patents—and you would be right. Over 22,000 U.S. patents related to LED technology have been granted, mostly in the past ten to fifteen years. Many companies, including Samsung, LG, Cree, Philips, GE, and Osram, are suing each other in numerous patent litigation fights. Can any legal system intelligently sort out patent rights among 22,000 patents in any reasonable time frame? (The question answers itself!)

As with all inventions, the ideas that went into the invention of the white-light LED are *derivative ideas*—ideas derived from thousands of previous ideas, in the case of the LED, going back over 100 years.

In 1907, in a letter to the editors of *Electrical World*, **H. J. Rounds** wrote that "On applying a potential of 10 volts between two points on a crystal of carborundum, the crystal gave out a yellowish light…but with 110 volts a large number could be found to glow…others gave instead of a yellow light green, orange or blue." The first scientific study of this phenomenon of luminescence in carborundum (silicon carbide) was carried out by the Russian technician **Oleg Vladimirovich Losev** in the 1920s and 1930s. Tragically, after publishing 43 research papers in Russian, German and British journals and being granted 16 patents, he died during the blockade of Leningrad in 1942 at the age of 39. Much of Losev's work was forgotten, and it wasn't until the 1950s, after the discovery of the transistor, that scientists at Bell Labs were able to explain the observed electroluminescence in silicon and germanium using the quantum theory of solid-state material.

Ruben Braunstein, working at RCA in 1955, generated infrared light from a gallium arsenide semiconductor diode. **Bob Baird** and **Gary Pittman** at Texas Instruments obtained similar results, and in 1962 filed for the first patent for an infrared LED, which was issued in 1966. **Nick Holonyak, Jr.**, working with **Dr. Robert Hall** at General Electric in the early 1960s, produced a red LED, the first in the visible spectrum. Red, yellow and green LEDs became popular in the 1970s, but producing a blue LED proved much more difficult. A blue LED, along with the red and the green, would be needed to produce white light. Thus, the search was on to produce a blue LED.

In 1968, scientists at RCA thought that gallium nitride might be the material to produce a blue LED. The project was assigned to a young scientist named **Herbert Maruska**, who spent over a year learning how to make gallium nitride crystals. Maruska continued to work on the project while earning his Ph.D. at Stanford University. He finally succeeded in creating a very dim blue LED in 1972. However, the priorities at RCA were changing, and the project to develop a blue LED was finally canceled in 1974.

Over the next two decades, three Japanese scientists, **Isamu Akasaki** and **Hiroshi Amano** at Nagoya University in Japan, and **Shuji Nakamura** at the University of California at Santa Barbara, picked up the quest to develop a high intensity blue LED. By the mid-1990s, they had succeeded in developing a high-efficiency blue LED, and all three were awarded the Nobel Prize in physics for 2014. Nakamura was able to demonstrate a white LED by including a red, green and blue LED chip in a single package. However, a much better approached turned out to use only a single blue LED, which excited red and green light from a phosphor coating on the inside of a transparent cover. Some of the blue light continued through the transparent cover, combining with the red and green light to produce white light. Thousands of scientists and engineers have improved this technique over the past two decades to produce the bright, highly efficient white light LEDs we have today.

The 2014 Nobel laureates were rewarded for the invention of an efficient blue LED. But it is clear that that quest spanned 50 years, involving thousands of scientists and engineers, many of whose contributions were critical to the commercial success of white LEDs. Singling out three individuals for recognition with the Nobel prize may be appropriate, but it would seem inappropriate to single out one of the over 22,000 LED patents for special recognition. But that did not stop a jury in 2015 from awarding Boston University over $13 million in a patent infringement suit against three Taiwan-based defendants, who were producing white LEDs. The patent was for a method of preparing gallium nitride single crystal films used for producing blue LEDs. The inventor was a Boston University professor, **Theodore Moustakas**, whose patent and work were not mentioned by the Nobel committee in their review of the history of the blue LED. Of course, the inappropriateness of universities applying for patents, discussed in Chapter 13, applies here as well. Asking twelve randomly-chosen citizens on a jury to decide the complicated technical issues surrounding the invention of the blue LED would seem to be asking too much.

Chapter 16

Who Invented the Internet?

> *The Internet is one of the greatest innovations ever launched, and even now has vast potential as a force for great good. And it is a source of pride to all of us that this progress was set in motion by two talented Americans. Our economy, our lives, and our world have all been enriched by the imagination and the efforts of Robert Kahn and Vinton Cerf.*
>
> President George W. Bush
> Remarks honoring Presidential Medal of
> Freedom recipients at the White House
> November 9, 2005

Most young people today take it for granted that they can use their smartphones to access information anywhere in the world. But the *Internet*, which allows that to happen by connecting millions of computers through countless different computer networks, was a monumental technological achievement involving thousands of scientists and engineers over many years.

Robert Kahn was born in Brooklyn and earned a bachelor's degree in electrical engineering from City College in New York in 1960. In 1964, the year Kahn got his Ph.D. from Princeton University and joined the electrical engineering department at MIT, computers were large mainframe devices that you programmed using punched cards, typically submitted for overnight batch processing. Kahn's work at MIT was theoretical, and to gain some practical experience, he took a leave of absence in 1966 and joined Bolt, Beranek and Newman, BB&N, a small architectural acoustic firm in Cambridge. There he was given the freedom to work on anything that interested him. He started working on networking, because he thought it was an interesting problem to get computers working with other computers.

At the Advanced Research Projects Agency, (ARPA), in the U.S. Department of Defense, **Lawrence Roberts** and **Robert Taylor** were putting together a computer networking project called ARPANET, which was to link together numerous computers at research centers across the country. This was an idea that was suggested in a 1963 memo by **J. C. R. Licklider**, who at the time was Director of Behavioral Sciences Command and Control Research at ARPA. In 1968, ARPA issued a Request for Quotation (RFQ), to produce a four-node network, where each node contained a *packet switch*, called an Interface Message Processor or IMP. The idea of dividing a message into a series of packets, which are sent independently over a network, called *packet switching*, was introduced by **Paul Baron** at Rand in 1964, and independently by **Donald Davies** of the National Physical Laboratory (NPL) in the United Kingdom. BB&N won the contract to produce this four-node network, and Robert Kahn was instrumental in designing and developing the IMPs for each node. The first IMP was delivered to **Prof. Leonard Kleinrock's** network lab at UCLA. By 1969, ARPANET had four nodes, connecting host computers at UCLA, Stanford Research Institute (SRI), the University of California Santa Barbara, and the University of Utah.

When Kahn left BB&N in 1972 to work at DARPA (ARPA had added a D, for Defense, in front of its title), the number of nodes in the ARPANET had grown to over two dozen at numerous university and government host computers across the country. This was the year that **Ray Tomlinson** at BB&N introduced electronic mail on the network, which quickly became the most popular use of the network.

Meanwhile, other types of networks were being developed, including ALOHAnet at the University of Hawaii, which was a wireless packet data network, and Ethernet at Xerox PARC, in which computers on the network communicated over a common coaxial cable. It quickly became apparent that it would be useful for computers on different networks to be able to communicate with each other. Thus, was born the idea of an *Internet*.

Robert Kahn, now at DARPA, started working on this problem, and together with **Vint Cerf**, who had received his Ph.D. in computer science at UCLA in 1972 and then joined the faculty at Stanford University, published a paper in 1974 describing a protocol for packet network intercommunication. The key was an open architecture that allowed a host on one network to communicate with a host on a different network through a gateway, using a standard protocol called Transmission Control Program/Internet Protocol, or TCP/IP. In 1977, ARPANET was successfully connected to two other networks through gateways.

More and more universities and government agencies wanted access to the Internet, and during the 1980s many started their own TCP/IP networks, including the Department of Energy, NASA, the National Science Foundation, and others. By 1990, ARPANET had broken off its military portions into separate networks, and the old ARPANET was decommissioned. The National Science Foundation's NSFNET gave many more universities access to the Internet.

On October 3, 2000, Vint Cerf hosted an Internet Society panel discussion on *Business Method Patents*. Before letting the panelists go at it, he made the following statement:

> *"Before I allow these gentlemen their five minutes, I am going to steal just a few myself. I want to mention something about patenting and the Internet and its origins. One of the things that is peculiar and interesting about the Internet history is that the TCP-IP protocols were never patented. In fact, they were made available as widely as possible to the public as soon as possible.... The openness of those protocols and their availability was key to their adoption and widespread use. I think if Bob and I had not done that—if we had tried to, in some way, constrain and restrict access to those protocols, some other protocol suite would probably be the one we'd be using today..."*

The TCP/IP protocols are a good example of the benefits that come by not having patents. Sadly, many of those that went on to add improvements to the Internet did not follow the good example set by Kahn and Cerf, adding hundreds of questionable patents to the pool, resulting in countless patent litigation suits, and unquestionably, slowing down the development of the Internet.

The demand for public access to the Internet increased, and during the 1990s several commercial Internet Service Providers started to fill this need. However, while scientists and engineers at research laboratories were willing to put up with awkward methods of accessing the Internet, the public would need a much simpler and intuitive method. That method would be born in a most unlikely place, a facility in Europe conducting atomic physics research.

Chapter 17

Who Invented the World Wide Web?

> *Friends at CERN gave me a hard time, saying it would never take off—especially since it yielded an acronym that was nine syllables long when spoken. Nonetheless, I decided to forge ahead. I would call my system the "World Wide Web."*
>
> Tim Berners-Lee
Weaving the Web
Page 23

Just as the Internet prospered by not having its basic protocols restricted by patents, the same commitment to open access would allow the *World Wide Web* to grow from the idea of a single software engineer, living near the France-Switzerland border in 1989, to the means by which millions of people would be able to access and share information worldwide a decade later.

Tim Berners-Lee was born in London, the son of two mathematicians who helped program the Manchester University "Mark I," the first commercial stored-program computer. From the time that he was in high school, he had the idea that computers could become more powerful if they could link unrelated information together. By the time he was working at the European Particle Physics Laboratory in Geneva, known as CERN, in the 1980s, the idea had jelled into a need to organize all of the disparate information that the scientists and engineers at CERN were generating. He wrote an internal proposal to create such a linked information system in March 1989, based on a personal information system he had been working on, off and on, for the past decade. When his proposal was ignored, he resubmitted it in May 1990, after which he got the go-ahead to proceed.

He called his system the *World Wide Web*, in which every piece of information in the web would have a unique identifier that he called a *universal resource identifier* or URI. This name would later be changed to *uniform resource locator*, or URL. To a user, or *client*, the computer screen would contain a page of text, some words of which would be highlighted. Selecting the highlighted word, perhaps by clicking a mouse, would bring up a new screen containing the information associated with that word. This method of accessing linked information in a non-sequential way is called *hypertext*, a term introduced by **Ted Nelson** in 1965. Berners–Lee wanted to combine hypertext with the Internet, allowing all of the information in the world to be linked. He defined a computer language called Hypertext Markup Language (HTML) and a protocol called Hypertext Transfer Protocol (HTTP) for transferring information across the web. By the end of 1990, he had a browser working on his computer, which could communicate with a server at CERN.

Over the next two years, Berners–Lee encouraged others to use his system. Students at many universities started writing browsers, including **Pei Wei** at the University of California Berkeley, who created a web browser in 1992 called ViolaWWW, which could display graphics with animations, and download small applications from the web. In April 1993, CERN put all web protocols and code in the public domain, allowing anyone to create browsers and servers free of charge, with no constraints.

At the National Center for Computing Applications (NCSA) at the University of Illinois at Urbana–Champaign, a student, **Marc Andreessen**, and a staff member, **Eric Bina**, created a browser they called Mosaic, which they made available over the web in February 1993. After graduating in 1993, Andreessen moved to California where he met **Jim Clark**, founder of Silicon Graphics. Together they founded Mosaic Communications Corporation to market their web browser. The University of Illinois didn't like them using the name Mosaic, so in 1994 they changed the name of the company to Netscape and the name of the browser to *Netscape Navigator*.

Netscape Navigator became wildly popular, in part, by giving it away for free. Netscape went public in 1995, the same year that **Bill Gates** and Microsoft, who were slow to recognize the importance of the Internet and the World Wide Web, introduced their own web browser called *Internet Explorer*. Understanding the threat from the giant Microsoft, Netscape took a bold move in January 1998 when it released the source code for Netscape Navigator to the public. This idea of open-source software, which has been extended to many different areas in the past two decades, has the advantage of allowing anyone to use and modify the source code. As a result, software bugs are fixed quickly, and the overall software quality is enhanced. In 1999, AOL purchased Netscape for $4.2 billion.

CERN and MIT formed the *World Wide Web Consortium* in 1994, and Berners–Lee moved to MIT to become its director. In fulfilling its goal to lead the web to its full potential, the consortium held periodic workshops for its members, recommending best practices, and releasing all software developed by the consortium to the public domain.

The fact that Berners–Lee had not patented any of the software related to the World Wide Web insured that the web should grow quickly, without hindrances. Surely, no one would want to start filing patents for small enhancements in browsers, and thus strangle advancements of the World Wide Web with needless patent infringement litigation. And you wouldn't expect universities to be involved in such mischief. But alas, the patent system almost invites the creation of patent trolls as a way of making a quick buck.

On October 17, 1994, **Michael D. Doyle** and others, on behalf of The University of California, San Francisco, filed for a patent entitled, *Distributed Hypermedia Method for Automatically Invoking External Application Providing Interaction and Display of Embedded Objects Within a Hypermedia Document*. This patent described web browser enhancements that Pei Wei had demonstrated two years earlier in his ViolaWWW web browser. Nonetheless, the patent office issued this patent, number 5,838,906, on November 17, 1998. In 1994, the year the patent was filed, the University of California, San Francisco, licensed the patent to the company Eolas, founded by Michael D. Doyle in Tyler, Texas. Microsoft refused an offer to license the technology from Eolas, and in 1999, Eolas sued Microsoft for patent infringement. In 2003, Microsoft lost in Federal Court in Chicago and was ordered to pay $521 million in damages. Alarmed by this court result, Berners–Lee, on behalf of the World Wide Web Consortium, provided evidence to the United States Patent and Trademark Office that the '906 patent was invalid due to prior art. After numerous appeals, Microsoft and Eolas settled the case for an undisclosed amount. In 2009, Eolas sued 22 other companies for infringing the same '906 patent. In a Texas jury trial in 2012, in which Tim Berners–Lee and Pei Wei testified on behalf of defendants Google, Yahoo and Amazon, the '906 patent was found to be invalid due to prior art, a finding that was upheld in 2013 by the U.S. Court of Appeals for the Federal District.

For nearly two decades, the advancement of the World Wide Web was threatened by a patent that never should have been granted in the first place. Microsoft paid millions of dollars to settle a case for an invalid patent. Countless companies wasted time devising workarounds to avoid infringing an invalid patent. Does this sound like a patent system that is fair?

Chapter 18

What Makes a Phone Mobile?

> *"Joel, this is Marty Cooper.... I'd like you to know that I'm calling you from a cellular phone....Only this is a real cellular phone, portable, handheld."*
>
> Martin Cooper, of Motorola, placing the first ever public call on a handheld cellular telephone from a Manhattan sidewalk to his competitor, Joel Engel, at Bell Labs. April 3, 1973

When I was a child growing up in the 1940s, the telephone at our summer farmhouse was a box mounted on the wall with a crank and a microphone you spoke into, and a removable cone on a cord that you held up to your ear to listen. We were on a party line with six other customers, any one of whom you could call directly by turning the crank to ring their number. Four short rings, and my aunt at the top of the hill would answer her phone. Our phone number was one long ring followed by two short rings. Anyone on the party line could listen in to anyone else's phone call. To call someone not on your party line, you would crank one long ring and be connected to Flossie, the local telephone operator, sitting at her switchboard in town.

The telephone at our winter home was a black phone that sat on the desk. To make a call, you would lift the phone from its cradle and tell the operator the number you wanted to call. Our number was 353-W. Almost all the calls we made were local calls, long-distance calls being much more expensive. To make a long-distance call, you had to ask for a long-distance operator, who would connect you to the city and number you wished to call. Eventually, our black phone and our crank phone were replaced by a dial phone, where we could dial the number directly, without talking to an operator. Later, our dial phone was replaced by a pushbutton phone. However, in all of these cases, the phone was located in a particular place in the house, connected by wires to the phone lines, which crisscrossed the town and country on telephone poles.

The idea of being able to make a telephone call without wires is an old idea—going back over 100 years. As mentioned in Chapter 8, Reginald Fessenden was interested in wireless telephony as early as 1900. In December 1906, he demonstrated a wireless telephone connection between Brant Rock and Plymouth, Massachusetts, a distance of eleven miles.

Two-way radios were being used by commercial and military ships by 1912 and began to be used in police cars in the 1920s and 1930s, allowing the cars to communicate with a central station. Two-way radios were used extensively by soldiers in World War II.

The problem of many people using a wireless phone is the same problem of many radio stations operating in the same area—how to keep them from interfering with one another. The radio solution is to assign different frequencies to different radio stations. What parts of the electromagnetic spectrum are used for different purposes is determined by the Federal Communications Commission (FCC). The low-frequency audio signals that are broadcast on radio must first be superimposed on a higher frequency carrier signal—a process called *modulation*. Traditional analog radio uses two types of modulation—*amplitude modulation* (AM) and *frequency modulation* (FM).

In amplitude modulation, the amplitude of the carrier signal is varied in proportion to the low-frequency audio signal, while in frequency modulation, it is the frequency of the carrier signal that is varied in proportion to the audio signal. The carrier frequencies for AM radio range from 530 kHz to 1710 kHz. To prevent interference, the frequencies assigned to the radio stations are separated by 10 kHz. This so-called *bandwidth* is twice the highest frequency of the audio signal, 5 kHz, which is high enough to transmit speech and low-fidelity music. This 10 kHz bandwidth per station means that the AM broadcasting band will accommodate 119 stations. The carrier frequencies for FM radio range from 88 MHz to 108 MHz, with each station using a typical bandwidth of 100 kHz or 200 kHz, accommodating either 100 or 200 radio stations.

Wireless phones would clearly need to accommodate more than 100 or 200 users at a time. What big ideas would make this possible? The first big idea was suggested in an internal Bell Laboratories technical memo written by **D. H. Ring** on 11 December 1947, entitled *Mobile Telephony—Wide Area Coverage*. In this memo, Ring outlines a mobile radio system in which vehicles equipped with such radio phones could operate throughout the entire country. The basic idea of such a system is to divide the country into small areas, and then to divide these small areas into even smaller areas called cells, arranged in a hexagonal layout, an arrangement suggested by fellow Bell engineer **W. R. Young**. A vehicle traveling through one of the cells would communicate with a station within that cell on a particular frequency. Adjacent cells would have stations operating at a different frequency. Distant cells could reuse the same frequency. The idea would be to keep the power of the radio phone low so as not to interfere with the same frequency at a distant cell. When the vehicle moved into an adjacent cell, it would automatically be switched to a new station within that new cell at a new frequency. This process would continue as the vehicle moved from one cell to the next.

Such a cellular phone system was ahead of its time in terms of the technology needed to implement it. It would be over 26 years later that **Martin Cooper** of Motorola made the first public cellular phone call quoted at the beginning of this chapter. In that year, 1973, the FCC assigned 30 MHz of the spectrum at 900 MHz for the purpose of developing demonstration cellular phone systems. It took another ten years to develop the first fledgling, commercial cellular phone systems. These were all analog phone systems, and the first such radio phones were large and awkward to use.

The next big idea in the development of cellular phones was to convert the analog audio signal into a digital signal—just a series of bits, ones and zeros—and to transmit this stream of ones and zeros instead of the analog signal. At the receiving end, this digital signal would be converted back into an analog audio signal. The rapid advancements in computer technology made this possible by the early 1990s.

With the advent of digital communications, it was possible to implement new ideas on how to modulate such digital signals onto high-frequency carrier signals. Traditional amplitude modulation and frequency modulation were easily adapted to signals with just two values, zero and one. *Phase modulation*, in which the phase, or position, of a sine wave is adjusted depending upon the value of the digital signal, also lends itself to digital communications. Combinations of these techniques, known as *quadrature amplitude modulation*, or QAM, are used to modulate strings of bits at one time, greatly increasing the speed of communication as well as reducing the bandwidth needed for each user.

One final big idea related to digital communications concerns how to pack more simultaneous users into the same band of the radio frequency spectrum. As we have seen, the way this was done for analog radio was to assign each user a different frequency, so-called *Frequency Division Multiplexing* (FDM). Digital signals can be split up into pieces, sending pieces of different user's signals at different times, called *Time Division Multiplexing* (TDM). However, a completely different, and counter-intuitive method is the one most likely used on your smartphone. Instead of trying to have each user use as little of the frequency spectrum as possible, the method is to have everyone use about 100 times the bandwidth as normal. But the secret is that everyone uses the same much wider bandwidth, called *spread spectrum*, at the same time, and it turns out that more users can be packed into the same wider bandwidth than if each user were given a much smaller individual bandwidth.

There are several different spread spectrum techniques—we will see a different one in the next chapter—but the one used in your cell phone is called *Code Division Multiple Access* (CDMA). In this method, each digital signal has its own specific code—a pseudo-random sequence of say 100 bits for each bit time of the original data signal. All the coded signals for all users are then added together and transmitted at the same time. At the receiving end, the specific code for each user can be used to extract the original signal for that particular user.

To see how this might happen, the *Exclusive-OR* operation can be used. The Exclusive-OR operation on two bits is zero (0) if both bits are the same (both 0 or both 1), and it is one (1) if both bits are different (one 0 and one 1). If you Exclusive-OR a bit with a 1, you will flip the bit—a 1 becomes a 0, and a 0 becomes a 1. On the other hand, if you Exclusive-OR a bit with a 0, the bit remains unchanged. Therefore, if you Exclusive-OR a 1 bit in the data signal with a code made up of 100 random 1's and 0's, all of the bits in the code will be changed. But, if you Exclusive-OR a 0 bit in the data signal with the same code, then all of the bits in the code will remain the same. This coded signal is the one that is transmitted. What will happen at the receiving end if you Exclusive-OR the received signal with the same code? For those data signal bits that were a 1, all of the coded bits will be opposite, and the result of the Exclusive-OR will always be 1. For those data signal bits that were a 0, all of the coded bits will be the same, and the result of the Exclusive-OR will always be 0. Thus, the original data signal will be recovered. On the other hand, a received signal from any other user will use a different pseudo-random code, and the Exclusive-OR operation will leave lots of 1's and 0's during the data signal bit time, which can be filtered out, leaving only the data from the user with the same transmitting and receiving code.

This idea of code division multiplexing was used extensively by the military starting in the 1950s, because such signals were more secure—the signal being undetectable without the proper code—and also more immune to jamming. For the same reasons, CDMA started to be used in military and commercial communication satellites in the 1970s and 1980s.

It should be clear that many big ideas and thousands of little ideas took decades to come together into the commercial cellular telephone systems we have today. Of course, this development was accompanied by the filing and granting of thousands of patents, together with the inevitable, endless patent litigation fights. The idea that anyone should be prohibited from building a better cell phone, because of one out of thousands of patents related to the cell phone, is ludicrous.

Chapter 19

What's with Wi-Fi and Bluetooth?

> *This innovation relates broadly to secret communication systems involving the use of carrier waves of different frequencies, and is especially useful in the remote control of the dirigible craft, such as torpedoes.*
>
> *An object of the invention is to provide a method of secret communication which is relatively simple and reliable in operation, but at the same time is difficult to discover or decipher.*
>
> Hedy Kiesler Markey
George Antheil
U.S. Patent 2,292,387
Filed: June 10, 1941
Issued: Aug. 11, 1942

The idea of jumping to different frequencies as part of a communication system was part of an invention for a secret communication system, patented by the inventor **Hedy Kiesler Markey** in 1942. This idea would surface fifty years later as a means of making Wi-Fi and Bluetooth work. Hedy Kiesler Markey was her maiden and married name in 1941, the year she filed for the patent. She was an Austrian-born actress, who met **Louis B. Mayer** in Paris in 1937. He persuaded her to change her name to **Hedy Lamar** and brought her to Hollywood the following year, where she became a movie star. From 1940 through 1942, which included the years her patent was filed and issued, she starred in nine motion pictures opposite leading men that included Spencer Tracy, Clark Gable, James Stewart, Robert Young, William Powell and Walter Pidgeon. In 1949, she starred opposite Victor Mature as *Delilah* in *Samson and Delilah*, a box office success, and her first film in Technicolor. Hedy Lamar's co-inventor was **George Antheil**, a Hollywood musician and composer. The technique described in the patent for jumping from one frequency to another employed a moving record strip of the type then used by player pianos. Ideas can come from anywhere!

Bluetooth, which is used to connect your smartphone to portable wireless speakers, including those in your car, and other types of Bluetooth–enabled devices, was invented by **Dr. Jaap Haertsen**, working at the Swedish company Ericsson in 1994. A Bluetooth *piconet*, meaning very small network, can contain between two and eight devices, one of which is a master and the others are slaves. Different piconets can overlap to form a *scatternet*. Communication interference between devices is eliminated by using a frequency hopping system, much like that envisioned by Hedy Lamar in her 1942 patent. In Bluetooth, there may be up to 78 channels, each 1 MHz wide, in the 2.4 GHz frequency band. The frequency used to transmit a packet of data between the master and a slave changes in a random pattern 1600 times per second, according to a procedure determined by the master.

The band of frequencies between 2.4 GHz and 2.4835 GHz is known as the ISM, or industrial, scientific, medical band. For example, your microwave oven operates in this frequency range. In 1985, the FCC released this band for general, low-power amateur and commercial use. Small wireless communication networks, such as Bluetooth and Wi-Fi took advantage of this.

The frequency hopping scheme used in Bluetooth is a spread spectrum method in which the frequencies used are spread over many different 1 MHz bandwidth channels. A different spread spectrum method is used in most Wi-Fi systems. It is called *orthogonal frequency division multiplexing*, or OFDM. In this method, instead of jumping from one frequency to another, the frequency band is divided into many subcarrier frequencies, and the data signal itself is broken up into pieces, with each piece assigned to a different frequency subcarrier. These subcarrier frequencies are designed to be multiples of a base frequency, which makes them *orthogonal*, meaning that when added together they won't interfere with each other. There can be a large number of subcarrier frequencies, meaning that the data sent in one of these subchannels will travel at a slower rate than if the original data were sent at a single frequency. This means that the bandwidth of each subchannel can be greatly reduced, packing in more subcarrier frequencies, resulting in a much higher overall data rate.

Bluetooth and Wi-Fi are both IEEE standards; Bluetooth is IEEE standard 802.15.1, issued in 2002 and 2005, and Wi-Fi is IEEE standard 802.11 (with many updates, indicated by suffixes such as *a*, *b*, *g*, *n*, *ac*), adopted originally in 1997. If a standard includes technology covered by a patent considered to be essential, the patent holder must agree to reasonable and nondiscriminatory (RAND) licensing agreements in order to have the technology included in the standard. However, such agreements don't prevent lengthy court challenges.

One such lengthy court proceedings resulted from the Commonwealth Scientific and Industrial Research Organization (CSIRO) in Australia suing Cisco Systems, Inc. for infringing their patent number 5,487,069, entitled *Wireless LAN*. This patent, which was filed on November 23, 1993 and issued on January 23, 1996, lists five inventors and contains 72 claims. The scientists at CSIRO had been working on problems in radio astronomy and realized that some of their same mathematical techniques could be applied to the problem of wireless signals in a local area network reflecting off objects and interfering with each other. When the '069 patent was issued in 1996, a group including some of the inventors founded a company called Radiata, Inc. to produce and sell wireless chips based on a licensing agreement with CSIRO. In 2001, Cisco acquired Radiata and continued to pay over $900,000 in royalties to CSIRO until 2007. In the meantime, CSIRO offered licenses to other Wi-Fi industry participants, eventually receiving over $400 million in royalties. When CSIRO and Cisco could not agree on a new royalty rate, CSIRO filed suit against Cisco on July 1, 2011. A four-day trial was held in District Court, beginning on February 3, 2014. Rejecting both parties' rates, the District Court came up with its own rate schedule and concluded that Cisco owed CSIRO $16,243,067. Cisco appealed, and on December 3, 2015 the United States Court of Appeals for the Federal Circuit found that the District Court had erred in its calculations, vacated the District Court's judgment and remanded for the District Court to revise its damages award.

In none of these lengthy court battles, was the validity of the underlying patent ever considered. The issue was only how to calculate the amount of the damages. However, the patent itself is a classic example of the flaws inherent in the patent system. It pieces together many different ideas and inventions, all conceived and invented by others, to achieve its goal. It uses orthogonal frequency division multiplexing, (OFDM), which goes back a very long time and was described in detail by **Robert W. Chang** in a Bell System Technical Journal article in 1966. Central to the patent is the use of the *Fast Fourier Transform*, which goes back to the mid-1960s (or even the early 1800s) and is well-known and widely used. The same can be said for all of the other components making up their system, including forward error correction, steerable antennas, and all of the digital logic. None of these multitude of previous inventors stands to gain anything when a jury in the Eastern District of Texas awards millions of dollars based on unrelated legal arguments.

It is even worse when the winning plaintiff has never invented anything. In 2015, another jury in the Eastern District of Texas awarded Rembrandt IP, a well-known patent troll, over $15 million from Samsung for their use of Bluetooth, claiming it infringed a patent issued to inventor **Gordon Bremer**. The patent at issue, *System and Method of Communication Using at Least Two Modulation Methods*, has nothing to do with Bluetooth. Bremer admitted on the stand that he did not invent Bluetooth. He generates patents that are never reduced to practice and just gives them to Rembrandt IP in an attempt to get juries to award damages for what should be frivolous lawsuits. It is another example of how the patent system has truly jumped the rails.

Chapter 20

How Does Your Smartphone Know Where It Is?

> *The invention described herein may be manufactured and used by or for the Government of the United States of America for governmental purposes without the payment of any royalties thereon or therefor.*
>
> First sentence of Patent No. 3,789,409,
> *Navigation System Using Satellites and Passive Ranging Techniques,*
> Inventor: Roger L. Easton.
> Filed Oct. 8, 1970, Issued Jan. 29, 1974.

Roger Easton spent his career at the U.S. Naval Research Laboratory (NRL), where he developed a satellite navigation system called TIMATION, an acronym for TIMe navigATION, in which the principles of operation are fundamentally identical to the current *Global Positioning System* (GPS). In 2006, he received the National Medal of Technology, and in 2010, he was inducted into the National Inventors Hall of Fame. He died in 2014 at the age of 93.

Easton's patent, cited at the beginning of this chapter, is only six pages long, containing three figures and six claims. The goal of the navigation system was to locate a ship on the ocean. If one knew the distance from the ship to two different satellites, the location of the ship on the Earth's surface could be calculated using triangulation. Finding the location of an object not on the Earth's surface, such as an airplane, would require three satellites. The distance to a fourth satellite is used in the current GPS system to correct for errors in clock synchronization.

Because signals from the satellite travel at the constant speed of light, Easton's first idea was to calculate the distance by measuring the time required for the signal to travel from the satellite to the ship and multiplying that time by the speed of light. This was not a new idea—the then current land-based LORAN (LOng RAnge Navigation) system also measured the time for signals to travel from transmitting towers. But a satellite-based system could cover a much wider area, potentially the whole earth.

Easton's patent describes a different method for measuring the time it takes for a signal to travel from the satellite to the ship, requiring high precision clocks on the satellites and ship. The satellite would transmit seven different signals at different frequencies, ranging from 100 Hz to 100 kHz, all derived from a single, stable 5 MHz clock signal. The receiver on the ship would generate these same seven clock signals and measure the phase shift due to the propagation delay of each of the seven signals sent from the satellite. The multiple signal frequencies allowed the location of the ship to be calculated more accurately. Think of a clock that ticks only once every hour. You would only know the time to the nearest hour. But if you could see a second clock that ticked once every minute, you would know the accuracy to the nearest minute. If you had a third clock that ticked once every second, you would then know the time to the nearest second. This is the basic idea of having the satellite send multiple frequency signals.

In the United States GPS system, 24 satellites circle the earth in six different orbital planes (four satellites per orbit) at an altitude of about 20,000 km. Each satellite completes two full orbits per day, and at least eight satellites are visible at the same time from any place on earth. The satellites can all transmit at the same carrier frequency and use **Code Division Multiple Access (CDMA)**, as described in Chapter 18, with each satellite using a different **pseudo-random code to prevent interference.** The receiver on the ground needs to know all of these pseudorandom codes in order to extract the signals from different satellites.

A satellite needs to send a lot of auxiliary information about its own elliptical orbit from which the receiver can calculate the position of the satellite. The data rate for sending this information is relatively low, and it typically takes a receiver about a minute to collect this information, calculate the satellites locations, and sync up with the satellites. Your smartphone probably doesn't take this long to get its GPS function going, because it uses an assisted GPS function in which it collects this satellite location information directly from a cellular tower using a normal cellular data channel.

If the receiver clock is not exactly synchronized with the clocks on the satellites, then all of the phase shift measurements will be off by a constant amount. By adding a measurement from a fourth satellite, this clock bias can be adjusted until any three of the four satellites produce the same location value. There are many other sources of errors, including effects of the ionosphere and troposphere and the problem of multiple reflections. Ingenious methods have been devised to correct for most of these errors. The GPS system on your smartphone is probably accurate to about 5 meters (16 feet), but new technology may soon increase that accuracy to under a foot. The current GPS system can actually be much more accurate than that by receiving two different carrier frequencies instead of just the one used in your smartphone, but such accuracies are mostly limited to specialized, high-end and more expensive systems.

When the GPS system was initially fully implemented in the early 1990s, it was used by the military in support of military operations. Pressure grew to allow its use for civilian applications, but additional errors were introduced to intentionally degrade the accuracy for civilian use. However, on May 1, 2000, in a press release from the White House, President Bill Clinton stated, "Today, I am pleased to announce that the United States will stop the intentional degradation of the Global Positioning System (GPS) signals available to the public beginning at midnight tonight." Doubts persisted among the public, and so seven years later, another White House press release stated,

> Today, the President accepted the recommendation of the Department of Defense to end procurement of Global Positioning System (GPS) satellites that have the capability to intentionally degrade the accuracy of civil signals. This decision reflects the United States strong commitment to users of GPS that this free global utility can be counted on to support peaceful civil activities around the world.

> This degradation capability, known as Selective Availability (SA), will no longer be present in GPS III satellites. Although the United States stopped the intentional degradation of GPS satellite signals in May 2000, this new action will result in the removal of SA capabilities, thereby eliminating a source of uncertainty in GPS performance that has been of concern to civil GPS users worldwide.

The accuracy of a mobile GPS receiver can be significantly improved by using what is called *differential GPS*, in which the accurate locations of fixed sites on earth are used to further reduce GPS errors. There are many such public and private augmentation systems, including ones operated by the U.S. Coast Guard and the Federal Aviation Administration (FAA).

The obvious worldwide uses of the GPS system makes it not surprising that hundreds, if not thousands, of GPS patents have been filed and issued, notwithstanding the fact that the entire system was developed with government funding. When such a system is developed and made available to the public, hundreds of obvious applications immediately suggest themselves. Does it make any sense to issue a patent to the first person to file for a patent on such an obvious application?

Chapter 21

Talking to Your Smartphone

> *Me*: Did the Red Sox win their game yesterday?
> *Siri*: The Red Sox snuck past the Phillies in thirteen innings yesterday; the final score was 2 to 1.
> Talking to my iPhone
July 31, 2018

To many teenagers, asking their smartphone a question such as the one shown above, and getting an immediate answer, seems routine; something they have always been able to do. But getting speech recognition by a computer to that point has taken thousands of researchers over fifty years of hard work.

In 1952, **K. H. Davis** and his colleagues at Bell Labs designed and built a speech recognition system called *Audrey*, which could recognize the ten digits 0 through 9. This was a completely analog system using vacuum tubes that was large and not practical for commercial use. By the 1960s, computers could be used to try to recognize speech.

Suppose you wanted to have the computer recognize just two words: *yes* and *no*. If you recorded these audio signals and looked at them on an oscilloscope, you would see a messy, noisy pattern, which would look different for the two different words. These speech signals would be made up of many different audio frequencies, typically between 250 Hz and about 4000 Hz. The word *no* would be expected to contain more low-frequency components, while the word *yes* would be expected to contain more high-frequency components. Therefore, the first step in almost all speech recognition systems is to determine how much of each different frequency is contained in the signal. You can calculate this so-called *frequency spectrum* by first sampling the signal, typically 20,000 times per second, and then calculating a *Discrete Fourier Transform* using the *Fast Fourier Transform* algorithm, published by **James Cooley** and **John Tukey** in 1965.

Once you have the frequency spectrum, you could divide it into two sections: a low-frequency section containing frequencies less than, say, 900 Hz, and a high-frequency section containing frequencies greater than 900 Hz. You now pick some feature associated with each section. For example, you could pick the maximum amplitude of frequencies in the low-frequency section as feature number 1 and take the maximum amplitude of frequencies in the high-frequency section as feature number 2. Or, you could pick the average magnitudes in each section; either one would probably work.

To train the system, you would say the word *no*, calculate the frequency spectrum and the values of the two features, and then plot the values of the two features on a two-dimensional graph where the horizontal axis is feature 1 and the vertical axis is feature 2. For the word *no*, with mostly low frequencies, you would expect the dot to be plotted in the lower-right corner, indicating a high value for low frequencies and a low value for high frequencies. You would then repeat the process by saying the word *yes*. This word should have mostly high frequencies and fewer low frequencies, so you would expect the dot to be located somewhere in the upper-left region of the graph. You would repeat this process many times, saying *yes* and *no* over and over again. You would expect a clump of dots corresponding to the word *no* to be grouped together in the lower-right corner of the graph, and you would expect the dots corresponding to the word *yes* to be grouped together in the upper-left region of the graph.

To recognize an unknown spoken word, you might first calculate the mean value of all the *no* dots on the graph and the mean value of all the *yes* dots. Then when you say a new word, the computer would simply calculate the location of the dot for the unknown word, and if it were closer to the mean value for *no* it would classify the word as *no*, otherwise, it would classify it as *yes*.

This is the basic idea of how most pattern recognition systems work. However, in the case of speech recognition there are many complications. If we were to modify the system we just designed to recognize the ten digits 0 through 9, then after training we would end up with ten different clumps of dots representing each digit, and most likely some of these clumps would overlap. Sophisticated statistical methods have been designed so that during the recognition phase, the most likely digit is selected. But what happens if our vocabulary is enlarged to say a thousand words? Now our two-dimensional plot becomes very crowded. We could add a third feature by dividing our frequency scale into three sections. Now our so-called *feature space* becomes a three-dimensional cube rather than a two-dimensional graph. We could continue adding features by dividing the frequency range into more and more sections, leading to a higher-dimensional feature space. This is what is typically done in speech recognition systems.

So far, we would expect our speech recognition system to only work well with the person who trained the system. For many years, this was the case. So-called *speaker-independent* speech recognition was very difficult, because everyone has different accents and ways of saying things. Such speaker-independent speech recognition systems have only become more accurate in recent years by the collection of a huge corpus of speech data representing a broad spectrum of individuals.

Recognizing the ten digits, spoken individually, is one thing. Recognizing continuous speech, such as, "Did the Red Sox win their game yesterday?" is quite another. To handle large vocabulary, continuous speech, most speech recognition systems don't recognize individual words, but rather recognize certain units of sound, called *phonemes*. These would include the sounds made by the letters *p*, *b*, *k*, *ch*, *g*, *sh*, *m*, *h*, and many others. Then, sophisticated statistical and deep learning algorithms are used to string the recognized phonemes together, and coupled with natural language processing algorithms, the most likely string of words are predicted. It took the huge increase in computer speed and the massive increase in storage capabilities of recent years to bring together all of the speech recognition research of the past 40 or 50 years to give us the amazing speech recognition systems we have today.

Much of the speech recognition research done in the past 40 or 50 years was carried out in academic institutions, mostly funded by government grants. When the possibilities and benefits of speaker independent speech recognition became apparent in the early 1990s, related patent filings increased exponentially, with over 20,000 filed and issued by 2015. The companies who collected large speaker recognition patent portfolios include Microsoft, Nuance, AT&T, IBM, Sony, and Google. Hundreds of related patent litigation cases have been brought, mostly by patent trolls, euphemistically called a non-practicing entity (NPE), meaning the suing company doesn't actually practice anything with the patent—they are only in business to make money from patent infringement judgments.

There are so many ideas originating in the 40 or 50 years of academic research in speech recognition, that no speech recognition patent could reasonably claim to have found the one little idea that unlocked the secret to making speech recognition viable. One typical speech recognition patent cited over 300 previous patents and over 500 academic journal articles of related work. Clearly, whatever new little twist to the problem is claimed in the patent pales in significance to all of the previous work upon which it rests. To ask judges and juries to find that one specific speech recognition patent somehow supersedes the other 20,000 and therefore deserves some large compensation, is to ask the impossible.

By the way, Apple is way down on the list of companies with large speech recognition patent portfolios. The technology for Siri is provided to Apple by Nuance. And incidentally, Siri's voice is the voice of a real person—**Susan Bennett**, who spent four hours per day for a month recording long scripts of text. She didn't do it for Apple but for a speech recording company. She didn't know she was Siri until a friend told her that Siri sounded like her.

Chapter 22

Keeping Your Information Secret

> *We stand today on the brink of a revolution in cryptography.... A private conversation can therefore be held between any two individuals regardless of whether they have ever communicated before. Each one sends messages to the other enciphered in the receiver's public enciphering key and deciphers the messages he receives using his own secret deciphering key.*
>
> Whitefield Diffie and Martin E. Hellman, "New Directions in Cryptography," IEEE Transactions on Information Theory, Vol. IT–22, No. 6, November 1976.

In 1974, **Martin Hellman** was a young professor at Stanford University, who had previously worked at the IBM research labs in Yorktown Heights, New York where he learned of the work they were doing on cryptography for the government, particularly the National Security Agency (NSA). This work at IBM led to the *Data Encryption Standard* (DES), proposed by the National Bureau of Standards (now named the National Institute of Standards and Technology—NIST) and supported by the NSA. This encryption method used a 56-bit key, the same key being used for both encryption and decryption. Such a system is called a *symmetric-key* encryption system.

Whitfield Diffie had been interested in cryptography since he was a child and graduated from MIT in 1965 with a degree in mathematics. In the summer of 1974, during a visit to the IBM research labs in Yorktown Heights, New York, he met the head of the IBM mathematics group, **Alan Konheim**, who couldn't tell Diffie anything about the work they were doing on cryptography for the government because it was secret but suggested that he talk with Martin Hellman at Stanford University. In what was supposed to be a half-hour meeting in Hellman's office turned into a two-hour meeting, followed by dinner at Hellman's home, with the discussions continuing till about 11 o'clock at night.

Diffie and Hellman both realized that the 56-bit key used in the DES encryption method was inadequate and thus insecure. Diffie knew that if the Internet was going to be used for online shopping and banking, as was being proposed, a better method of securing data would be needed. In symmetric encryption methods, such as DES, the same key is used for encryption and decryption. Getting the key to the receiver traditionally required some type of secure courier service. This would clearly be impractical for thousands of transactions taking place over the Internet. Some different method would be needed.

In the spring of 1975, Diffie came up with the concept of a digital signature, which would identify a particular person. He then realized that the same idea could be extended to sending encrypted messages without first having to send the key by other means. The basic idea is to have two keys—a public key and a private key. Anyone can see the public key, but the private key must be kept private by a particular user.

To understand how such a public key encryption system works, suppose Alice wants to send Bob a secret message. (All discussions of public-key encryption use Alice and Bob as the people sending messages to each other. The person eavesdropping is called Eve.) First Bob sends Alice his public key. Because it is public, he doesn't care if anyone else sees it. Alice puts her secret message in a box and locks the box by sticking Bob's public key in the keyhole and turning the key. But Bob's public key is a magic key, and when Alice removes Bob's public key from the keyhole, the inside structure of the keyhole changes so that now only Bob's private key can unlock the box. Alice then sends the box to Bob, and if Eve tries to intercept the message, she will not be able to unlock the box because she does not have Bob's private key. When Bob receives the box, he simply unlocks it with his private key and removes Alice's secret message.

Bob's public and private key are clearly related somehow, because his public key must know how to modify the keyhole such that only his private key can then unlock the box. So, the question is, can someone look at his public key and figure out how the keyhole is modified and thus learn what his private key is? In practice, the public and private keys are long strings of bits, and the keyhole and the process of unlocking the box are sophisticated mathematical functions and operations, such that locking and unlocking the box with the proper key is easy but guessing the private key, knowing only the public key, is very difficult—so difficult, that trying to do it by brute force would take millions of years.

Now suppose that Alice puts a message in her box and locks the box with her *private* key. When she removes her private key from the keyhole, the internal structure of the keyhole magically changes so that only her public key can unlock the box. But now anyone knowing her public key can unlock the box and see the message, so you might think this is useless. But actually, it is very useful. If someone unlocks her box and removes the message, which says, "I'm Alice," they will know that the message had to come from Alice, because only Alice's private key could have locked the box. This technique is used today to authenticate the identity of a particular user.

As a young computer scientist at MIT in 1977, **Ronald Rivest** read with interest Diffie and Hellmann's paper, "New Directions in Cryptography," cited at the beginning of this chapter. Together with his colleagues, **Adi Shamir** and **Leonard Adleman**, they tried to come up with a scheme that would make such a public-key cryptographic system practical to implement and unbreakable. They tried many ideas before settling on a method that involved multiplying two large prime numbers together. The product of these two large prime numbers became part of the public key but finding the private key would require factoring this large product into its two prime factors, which is a very difficult task for large numbers. Based on the first initials of their last names, their method became known as the RSA public-key cryptographic method.

In the August 1977 issue of *Scientific American*, **Martin Gardner** wrote a column about the RSA public-key cryptographic method in which Rivest, Shamir and Adleman offered $100 to anyone who could factor a particular 129-digit number into its two prime factors and decode an encrypted message. The problem was solved in 1994 using thousands of computers on the internet.

Prior to 1985, pure mathematicians had been studying elliptic curves, an obscure topic in number theory, for 100 years. But in 1985, **Victor Miller** at IBM and **Neil Koblitz** at the University of Washington independently proposed that elliptic curves could be used for public-key cryptography. Since that time, the field of elliptic curve cryptography has exploded, and today it is the preferred method for public-key cryptography. The main reason for this is that stronger encryption can be achieved with shorter keys, making the calculations faster while using less memory. This makes it practical to use elliptic curve cryptography with mobile wireless devices and even smart cards. For example, a 571-bit elliptic curve cryptographic system would be as difficult to break as a 15,360-bit RSA system, while running 400 times faster.

It turns out to be impractical to use public-key encryption methods to encrypt an entire message. Rather, a public-key encryption method is used to encrypt a symmetric key, which is then used with some type of standard symmetric-key encryption method.

Diffie and Hellman first presented their public-key cryptographic system at a conference in June 1976. It turns out that Hellman himself presented the paper, rather than one of his graduate students as was his usual practice, because the NSA had claimed that discussing their method publicly would violate U.S. law that prohibited the export of weapons to other countries. The lawyers at Stanford did not agree with this interpretation of the law but felt that it would be easier to defend Hellman rather than one of his graduate students. In the end, he gave the paper, and no one was arrested.

Diffie and Hellman's IEEE publication cited at the beginning of this chapter was published in November 1976. On September 6, 1977, they filed for a patent (including the name **Ralph C. Merkle**) on *Cryptographic Apparatus and Method*. This filing date was over a year after they presented their conference paper in June 1976, which by patent law should have made the patent invalid. Nonetheless, their patent was issued on April 29, 1980.

Meanwhile, on December 14, 1977, Rivest, Shamir and Adleman filed for a patent on *Cryptographic Communications System and Method*, which was issued on September 20, 1983. In 1982, Rivest, Shamir and Adleman founded RSA Data Security to commercialize the RSA public-key cryptographic system. Their patent expired in 2000, at which time RSA put all of their methods in the public domain.

The Diffie and Hellman patent was assigned to Stanford University, and the Rivest, Shamir and Adleman patent was assigned to MIT. All of the discussion we had in Chapter 13 regarding the Google patent and the inappropriateness of patents being filed by academic institutions holds in the case of these two patents as well. The development of the public-key cryptographic system is a good example of how academic research is done—lots of discussions with lots of people, going down many blind alleys before a solution is found. And once a solution is found, you are often surprised to learn that someone else had thought of the same solution long before you did.

In fact, in 1970, **James H. Ellis**, who was working in the Government Communication Headquarters (GCHQ) in England, published an internal report entitled "The Possibility of Secure Non-Secret Digital Encryption," in which he outlined the public-key encryption method six years before Diffie and Hellmann's publication. After joining GCHQ in 1973, **Clifford Cocks** was told about Ellis's method, but that no one knew how to implement it. After thinking about the problem for a while Cocks came up with essentially the RSA method. However, all of this work was classified, and when Diffie and Hellman published their paper in 1976, only the NSA knew that this idea had already been conceived years earlier in England. Computer speeds at the time made the method impractical, so the pioneering work in public-key cryptography at GCHQ remained classified until 1997, when on December 18th Clifford Cocks gave a public talk describing this early work. James H. Ellis had died a month earlier.

Chapter 23

Shopping on Your Smartphone

> We claim:
> 1. A method of placing an order for an item comprising: under control of a client system,
> > displaying information identifying the item; and in response to only a single action being performed,
> > > sending a request to order the item along with an identifier of a purchaser of the item to a server system;
>
> > From claim 1 of Patent No. 5,960,411,
> > *Method and System for Placing a Purchase Order Via a Communications Network,*
> > Assignee: Amazon.com, Inc.
> > Filed Sept. 12, 1997, Issued Sept. 28, 1999.

Why would the patent office grant a patent to Amazon for its 1-Click checkout operation on its website? There's nothing non-obvious about how to implement such an operation—any competent computer science graduate could implement such a system if asked to do so. There is nothing new about collecting credit card information ahead of time, such as at a Country Club, so that every time a charge needs to be made, it can be done by simply identifying oneself. The only thing new was applying this obvious idea to buying things on the Internet. Apparently, the newness of the Internet blinded patent examiners to something so obvious.

There was much controversy about granting this patent, which was described as a "business method" patent. But do such business method patents "promote the progress of science and the useful arts?" Does it make sense for one company to have a twenty-year monopoly on using any business method? Whether any business method is useful or not should be determined in the marketplace, not in a patent infringement court of law.

In October 1999, Amazon brought a patent infringement suit against Barnes & Noble for their "Express Lane" one-click checkout option. The case was settled in 2002, but not before Barnes & Noble added a second "confirm" click to get around the patent. Up until the time the patent expired on

September 12, 2017, **Jeff Bezos** continued to defend the patent, although he suggested that perhaps such business method patents should expire in four or five years. (How about four or five minutes!)

As discussed in the previous chapter, the ability to encrypt your data on the Internet and keep it secret has led to the trust that made online shopping possible. Online banking is routine today. Millions of people file their income taxes electronically. Doctor's offices have portals where you can review your medical records. Ordering and purchasing prescription drugs online is routine.

Of course, the cost of some prescription drugs can give you sticker shock. There doesn't seem to be any relationship between the cost of one small pill and the cost of another small pill, which might cost ten times, a hundred times, or even a thousand times more. But of course, there is a reason. You guessed it—patents.

So far in this book, we have seen how the patent system has been abused and have suggested that abolishing the patent system could very well make things better. But the pharmaceutical industry has always been pointed to as the poster child of where the patent system is needed and works. It is argued that the research and development costs of bringing a new drug to market are extremely high, upwards of $2 billion, and that the risks of failure, late in the development process are also high. This, coupled with the ease of reverse engineering the product by a competitor, makes it essential to let a pharmaceutical company patent the drug and thus get a twenty-year exclusive right to produce the drug without competition. The fact that the FDA approval process might eat up 10 to 15 years of their exclusive 20 years, makes it all the more necessary to have at least these remaining five years of exclusivity.

That's the argument, and almost everyone seems to agree with it. The pharmaceutical industry spends at least $200 million per year on lobbying to make sure that everyone in Congress agrees with the argument. But is the argument really true? Would abolishing the patent system really demolish the pharmaceutical industry? Or, contrary to all conventional wisdom, would the pharmaceutical industry actually flourish? The reason that most people agree with the argument is that they haven't thought through what it would be like in a world without patents.

Chapter 24

A World Without Patents

> *There is freedom waiting for you,*
> *On the breezes of the sky,*
> *And you ask, "What if I fall?"*
> *Oh but my darling,*
> *What if you fly?*
>
> — Erin Hanson

By now, you may have thought of many reasons why the current patent system should be abolished. Here is a possible list:

1. *Ideas create wealth, and ideas are not patentable.*
2. *Scientific discoveries, upon which all inventions are based, are not patentable.*
3. *Patents are not the reason people invent.*
4. *Patents may not be awarded to the true inventor.*
5. *Patents stifle economic growth.*
6. *Turning ideas into reality often takes a long time.*
7. *Time and money are wasted litigating patent claims.*
8. *Judges and juries are not competent to judge the validity of patents.*
9. *Inventions always include the inventions of others.*
10. *Patents are monopolies, often requiring anti-trust actions.*

But the most important reason, inherent in all of these reasons, is that *the current patent system is not fair.*

It is not fair that a patent does not recognize all the ideas of others that make any invention possible. It is not fair that patent trolls tie up a court system for years, trying to cash in on inventions they did not make. It is not fair that the patent system stifles economic growth, contrary to its stated purpose. It is not fair that patents are sometimes not awarded to the true inventor. It is not fair that most patents could probably not pass a rigorous test of patentability. It is not fair that patent litigation decisions are made by judges and juries, generally unqualified to understand the technical issues in a case. It is not fair that companies find it less expensive to settle a dubious case, then to endure the time and expense of costly litigation. It is not fair that the public has to pay the hidden costs, in the form of higher prices, of constant patent infringement cases.

There must be a fundamental flaw in the patent system, leading to all of this unfairness. There is. *The legal system, in particular the patent system, is incompatible with technological innovation.* The patent system, contrary to its stated purpose, does not fuel innovation, but rather serves as a restraint on innovation. If one reads lots of patents and transcripts of patent infringement cases, one is struck by how little this massive legal structure has to do with how innovation actually occurs. The whole purpose of the patent system is supposed to encourage innovation. But the way the patent system has developed, it does just the opposite. Instead of latching onto a good idea and running with it, engineers and other innovators must always be looking over their shoulders, wondering if there is some obscure patent they might be infringing.

As pointed out in Chapter 1, wealth is created by ideas, lots of ideas, one building upon another, along with the freedom to develop these ideas, unhindered by unnecessary legal constraints. A patent is supposed to describe some invention, something so new and novel that it is worth preventing anyone other than the inventor from producing or profiting from the invention for twenty years. Recall from Chapter 12 that the brief brought by the fifty intellectual property professors in support of Samsung referred to an estimate of 250,000 patents related to the smartphone, an indication that everything these days, no matter how minor, is patented. Clearly, no court system can efficiently handle 250,000 patents. Something is fundamentally wrong. That something is a misunderstanding of how innovation and invention actually occurs.

A new innovation or invention begins with an idea. A new scientific theory begins with an idea. All scientific and technological advances are based on ideas. But you cannot patent an idea. The patent law states that you can only patent "any new and useful process, machine, manufacture, or composition of matter, or any new and useful improvement thereof." When the patent law was written 250 years ago, it was expected that the inventor would bring to the patent office the actual device representing the invention. However, today unimpressive software algorithms—pure ideas—are granted patents.

The basic problem with a patent is that when you drill down you find that all patents are based on ideas, mostly the ideas of others upon which the new invention rests. The only part of the patent that matters is the claims section at the end of the patent. A patent attorney will try to make these claims as broad as possible, by repeating the same claim over and over again with a slight modification. But in almost all cases, the invention hinges on one or two ideas designed to improve some previous invention, based on ideas which, in turn, improved some earlier invention. Only this new little idea is the thing that is patented—the thing that you can prevent others from using. In the case of the smartphone, there may be 250,000 such small ideas, and some patent holder is trying to convince some judge to halt the production of the entire smartphone, or else pay millions of dollars for using this one small idea that is embedded deep within the smartphone. But ideas are not supposed to be patented in the first place.

Much patent litigation has been brought on by companies who never produced anything. They simply buy up thousands of patents, most of which contain dubious claims if scrutinized carefully, and then essentially blackmail large companies into settlements, rather than wasting larger sums of money on lengthy litigations. Is this any way to promote innovation?

Another requirement for something to be patented is that the invention is novel and not obvious to someone skilled in the art. This requirement is something that is almost impossible for a judge or jury to determine. An attorney may argue that it must be nonobvious because no one has patented it before. But that same argument could be used to prove that it *is* obvious—so obvious to one skilled in the art that it would seem ridiculous to claim that it was novel. The patent office doesn't seem to appreciate the irony of awarding a patent to the inventor who files first. When multiple inventors independently invent the same thing, the invention can't be that nonobvious.

The other requirements for a patent is that the invention must be useful. But in most cases, this is something that can't be known at the time the patent is filed. Ultimately, only the market can determine the usefulness of an invention, and often this takes years to realize. The Motorola patent on the cell phone was filed in October 1973 and issued in September 1975. But it would take ten years and many new ideas and innovations before the first commercial cell phone system was operational. Holding up any single patent as being the critical key that makes something as complicated as the cell phone system—and later the smartphone—possible, simply ignores the way technological advances actually occur. Fundamentally, the patent system has nothing to do with it.

The purpose of the patent system "to promote the Progress of Science and useful Arts" is also fundamentally flawed. The current patent system is not what motivated all the engineers and scientists who created your smartphone. As pointed out in Chapter 1, they are having fun while getting paid to create new things. It's not the people doing the actual inventing who like patents, it's the lawyers and some business types who think that they can stifle the competition by tying them up in lawsuits. They buy other companies just to get their patent portfolio to use against their competitors. All of this wasted time and money would be better spent developing and selling new products.

Picture a world without the patent system. Engineers and scientists could work as they did in graduate school—publishing their results as journal articles or technical reports in a form that other intelligent people can understand, not as an unintelligible patent that no one will ever read. Best of all, they will be able to read what others have discovered and developed, talk to colleagues at conferences about new ideas, and use this information to speed up their own new development work. The net result is that everyone moves faster, not distracted by filing patents and trying to avoid interfering with the patents of others. The best ideas automatically rise to the top, adding to overall productivity and economic growth.

But the government already knows this, witness the cancelling of patent litigation fights during World War I, resulting in the accelerated development of radio. Recall that not patenting the World Wide Web or the TCP/IP protocol was a major reason for the rapid growth of the Internet.

Occasionally, an important discovery is made that changes the whole course of an industry. But even in those cases, patents are unnecessary. The discovery of the transistor is a good case in point. Bell Labs thought they could make a lot of money from this patent because everyone would need it. But the government will always break up monopolies, so in this case Bell Labs was required to license the transistor patent to any U.S. company free of charge. Recall that Gordon Moore stated that this was a primary reason for the growth of Silicon Valley. Eliminating the patent system altogether would simply mean that accelerated technological development would occur permanently in all areas. More Silicon Valleys would grow and thrive. Anemic economic growth is not inevitable.

Engineering professional societies develop all kinds of standards so that, for example, you can communicate over the internet seamlessly using equipment from any manufacturer. A complicated system has been established, involving government input, to ensure that "reasonable" license agreements are established with any patents that are essential to the standard. All of this extra work and resulting legal litigation goes away when the patent system is abolished.

Today, a patent expires in twenty years, after which anyone is free to use the invention. At the time that Edison and Westinghouse were fighting in court about the validity of Edison's light bulb patent, the length of a patent was seventeen years. This court litigation dragged out for years, with multiple appeals all the way to the Supreme Court. By the time that Edison won the case, there was only about a year left on the patent, and Westinghouse, in any event, had moved on to a different, non-competing design. This is typical of patent court cases—lots of wasted time and money, never really preventing a competitor from competing and developing a better product.

Abolishing the patent system would not eliminate all legal protections. Contract law would still apply to transactions between companies, and companies could use trade secrets as they do today. Most current software companies do not rely on patents. They may or may not keep their source code secret, but the success of any software company relies more on providing benefits to their customers than on trying to prevent competition by legal maneuverings. More and more companies are recognizing the benefits of *open source software*, where anyone can modify and de-bug the source code, resulting in much more reliable software. The Linux operating system is a good example, becoming the most widely-used open source software in the world. The top contributor to the Linux project is a company called Red Hat, which was bought by IBM in 2018 for $34 billion, an indication that open source software can be very profitable.

Small and middle-sized companies don't need patent protection to prevent larger companies from "stealing" their "inventions" and making a particular part, component or piece of software themselves. Even though, in principle, a large company could make anything it wants, in most cases, it is cheaper, and they will get a better product, by buying it from someone else. The free flow of ideas and products is the best way for all companies to grow and flourish.

Would abolishing the patent system be difficult to do? It takes changing only one word in the patent law. Change "twenty years" to "twenty minutes" for the length of time a patent is valid, and no one will ever file another patent application again!

If this is too much for Congress to do, there is another way. Those who do the inventing—the scientists and engineers—can simply stop filing patent applications. Publishing results and giving technical talks at professional meetings will put everything in the public domain. Everyone will have access to everything. No more wasted time in meaningless patent litigation. The rate of technological innovation will increase dramatically, resulting in a corresponding growth of the economy.

But how will scientists and engineers stop filing for patents? The way to do this is to flip the incentives. Instead of giving an inventor a chance of making money by excluding others from using their invention, as the present patent system does, give the inventor a chance of making money by allowing everyone to freely use their invention! How can this be possible?

Picture a completely new system—let's call it Open Science, Engineering and Technology, or Open-SET—where scientists and engineers publish the results of their research and development, including their "inventions," on an open website, with no gate-keepers deciding what is "suitable" for publication. Everything is accepted in a variety of formats—HTML, pdf, PowerPoint, video clips, articles, monographs, books, technical reports. Hyperlinks would allow easy access from one entry to another. The system would have easy access to existing online publications, including all patents in the current patent system. The user would be able to retrieve all of this vast technical information in a variety of useful ways, for example, grouping together all known knowledge of a particular topic.

Everyone would be able to use and build upon all previous contributions, writing their own improvements, or writing useful reviews about a particular topic. Everyone would be able to rate each contribution on a 5-star rating system, with comments. The popularity of each contribution would be recorded by posting the number of hits, as is done, for example, on YouTube.

Such a system would be so useful, that users would be willing to support it through a modest subscription. The number of subscribers would be so large as to generate substantial revenues. A large proportion of these revenues would be returned to the contributors in the form of "royalties," the largest royalties going to those with the greatest contributions in terms of ratings and hits. Such a system would have the proper incentives to "promote the progress of science and the useful arts."

Which brings us back to patents in the pharmaceutical industry. Would such a world without patents benefit that industry? One of the arguments for pharmaceutical patents is the high cost of R&D. But without patents, all research results from all sources would be immediately available for everyone to use. No longer would one be inclined to dismiss results from other sources, perhaps even old results, because they could not be patented. By leveraging research results from all sources, research costs would decline, and overall progress would accelerate throughout the industry—just as happened with the development of radio when patent litigations were suspended during World War I.

All this unrestrained research activity would accelerate research in newer areas such as Biologics, bioelectronics, and implantables. Machine learning and other artificial intelligence techniques are already being used to accelerate the discovery of new drugs. Does it make any sense for one computer program to have a patent over another computer program?

Another driver of high costs for prescription drugs is the long FDA approval process. As part of this process, the FDA can already grant a company some exclusive rights to market a drug, independent of that granted by a patent. Without a patent system, clinical trials could be broadened, involving multiple companies, perhaps supervised by a university, greatly reducing the time required to approve a drug. But clinical trials are really blunt instruments where the same pill is given to a large group of different people, all with different DNA. Even after a drug has been approved, there is no guarantee that that particular drug is the proper drug for a particular person. The long list of side effects accompanying most drugs attests to the blunt nature of today's clinical trials. How much better it would be if the accelerated research brought about by abolishing the patent system could result in a drug being tailored to a particular person's DNA, avoiding all potentially dangerous side effects. Ultimately, clinical trials could be a thing of the past. It used to be that to publish a book, the pages were transferred to large metal sheets, and a huge machine would crank out books on a large assembly line. Printing 3000 books at a time would be a typical, minimum run. Today, books are often printed one at a time when ordered. Will the day come when pills are not manufactured by the millions on an assembly line, but rather one bottle at a time, with the drug specifically tailored to an individual's DNA? This day certainly won't come if the pharmaceutical industry is restrained by a patent system that requires them to come up with a blockbuster drug periodically, on a twenty-year cycle, in order to survive. Will pharmaceutical companies become the next Kodak, unable to change in the face of new technologies, because they are too invested in their obsolete assembly lines?

The pressures for lower prescription drug costs are causing governments around the world to institute price controls. The reaction of the pharmaceutical companies is to charge much higher prices to those who can afford it, abandoning the needs of poor people in many countries. Such a situation is unsustainable, both economically and morally. Political pressures will always result in some government action. The 1984 Hatch–Waxman Act was one such reaction. This Drug Price Competition and Patent Restoration Act essentially established the generic drug industry in exchange for letting the pharmaceutical companies keep their 20-year patent exclusive rights. It is true that the price of a generic drug drops precipitously once a drug patent has expired, but the fact that the price of prescription drugs has increased by 69% since the year 2000, while the cost of physician and clinical services has increased by only 23%, means that drug price competition has a long way to go.

Do patents really provide the incentive for invention and innovation? Less than 5% of all patents are assigned to the actual inventor. The vast majority of patents are assigned to the inventor's employer—usually a company. The company owns the patent and any financial benefit goes to the company, not the inventor. The inventor may reap financial rewards through bonuses and stock options, which are also available to other employees. So, the real question is, do companies, pharmaceutical companies in particular, need patents in order to survive?

Since many companies, large and small, do survive without patents, we know it is possible. The argument that a pharmaceutical company needs a 20-year monopoly to offset the high research and development costs of a particular drug is flawed for several reasons. The company sells the drug for a very high cost, pricing many potential users out of the market. Many foreign countries have already imposed price controls on prescription drugs, banning many high-priced (and potentially life-saving) drugs from their countries. This is an indefensible state of affairs. The only real solution to the problem of the high costs of prescription drugs is not price controls, but competition. And the only way to get competition is to abolish monopolies, i.e., abolish patents.

But, if patents are abolished, will pharmaceutical companies stop doing research and stop producing new, life-saving drugs? It turns out that the majority of new drug developments are being done by start-up companies, most of whom are not profitable. Without patents, the research being done by all companies and universities would be immediately available for everyone to use in developing new drugs. This leveraging of research done by everyone would lower the research costs of any individual company—to say nothing of eliminating all of their patent litigation costs.

Wouldn't everyone then just produce the same drug? Not necessarily. Specialization would inevitably occur, as it does now—some companies specializing in cancer drugs, for example, others in drugs for treating heart disease. Some drugs may be sold by lots of companies. Many companies seem to make money selling aspirin and antibiotics today—without the aid of any patent.

The pharmaceutical industry is different from the smartphone industry in that it deals with people's health and lives. Because of this, it's important to get the incentives correct. It doesn't help if a patient can't afford a prescription drug. It doesn't help a patient when a company wants to continue selling a drug while they have exclusive patent rights, when a newer, better drug may be available. When a drug that cures a disease is less valuable to a company than a drug that the patient needs to take for the rest of his or her life, then the incentives are backwards.

Abolishing the patent system would begin to reestablish the proper incentives. No longer would an arbitrary 20-year drug cycle make any sense. The goal could shift to developing drugs to cure a disease rather than control it. New technologies that accelerated the discovery, testing and manufacturing of new drugs could be explored in an open environment where worries about patent infringements were a thing of the past.

A world without patents could be a wonderful and exciting world indeed!

www.ingramcontent.com/pod-product-compliance
Lightning Source LLC
Chambersburg PA
CBHW071200240526
45470CB00017B/757